The Enlightened Mr. Parkinson

The Enlightened Mr. Parkinson

The Pioneering Life of a Forgotten Surgeon and the Mysterious Disease That Bears His Name

Cherry Lewis

PEGASUS BOOKS
NEW YORK LONDON

THE ENLIGHTENED MR. PARKINSON

Pegasus Books Ltd
148 West 37th Street, 13th Floor
New York, NY 10018

First Pegasus Books hardcover edition August 2017

ISBN: 978-1-68177-454-1

10 9 8 7 6 5 4 3 2 1

Printed in the United States of America
Distributed by W. W. Norton & Company, Inc.

CONTENTS

John PARKINSON 1725–10 JAN 1784
m. Mary SEDGWICK 16 APR 1721–6 APR 1811

Elizabeth PARKINSON
26 JUL 1753–30 NOV 1754

James PARKINSON
11 APR 1755–21 DEC 1824
m. (17 MAY 1781) Mary DALE
2 SEP 1757–28 MAR 1838

Elizabeth PARKINSON
5 APR 1756–29 MAR 1759

James John PARKINSON
11 FEB 1783–3 JAN 1795

John W. Keys PARKINSON
11 JUL 1785–5 APR 1838
m. 1 (1811) Diana CHAPPLE
11 AUG 1787–FEB 1820
m. 2 (1821) Eliza TOMPSON
1800–3 MAY 1864

1 James Keys PARKINSON
27 OCT 1812–MAR 1849

1 Mary Hester PARKINSON
30 MAR 1816–8 FEB 1892

1 John Chapple PARKINSON
FEB 1820–MAY 1820

2 John Ellis PARKINSON
29 JUL 1822–1899

2 Elizabeth C. PARKINSON
2 JAN 1824–12 JAN 1900

2 Jane Dale T. PARKINSON
18 MAR 1825–10 DEC 1911

2 Emma T. PARKINSON
22 SEP 1826–26 NOV 1918

2 Caroline H. PARKINSON
27 FEB 1831–1839

2 Fanny S. PARKINSON
27 AUG 1834–8 OCT 1908

Emma Rook PARKINSON
5 APR 1788–27 AUG 1867
m. (1816) John DIMOCK
1791–8 SEP 1861

Emma P.W.K. DIMOCK
NOV 1821–3 APR 1906

Elizabeth C. DIMOCK
2 JAN 1824–12 JAN 1900

Jane Dale T. DIMOCK
18 MAR 1825–10 DEC 1911

Emma T. DIMOCK
22 SEP 1826–26 NOV 1918

Margaret Townley PARKINSON
3 AUG 1759–?1785
m. (25 JUL 1779) Richard WAYMAN
(dates unknown)
Children unknown

William PARKINSON
5 MAR 1761–29 AUG 1766

Mary Sedgwick PARKINSON
11 JAN 1763–APR 1840
m. (1784) John KEYS
1762–10 JUN 1836
No children

Jane Dale PARKINSON
2 OCT 1789–21 JAN 1792

Henry Williams PARKINSON
7 MAY 1791–APR 1862
m. (1820) Elizabeth ASPERNE
c. 1799–26 DEC 1875
Children unknown

Wakelin PARKINSON
17 OCT 1792–31 JAN 1793

Mary Dale PARKINSON
8 MAR 1794–16 MAY 1871
m. (1828) William TRINGHAM
(dates unknown)

Emma TRINGHAM
(dates unknown)

Caroline TRINGHAM
(dates unknown)

James TRINGHAM
(dates unknown)

Jane TRINGHAM
(dates unknown)

Mary TRINGHAM
1830–21 MAR 1914

William TRINGHAM
c. 1833–1909

Marianne TRINGHAM
(dates unknown)

PARKINSON
family tree

PROLOGUE

A hole in the head

English born and bred . . . forgotten by the English and the world at large – such is the fate of James Parkinson.

Leonard Rowntree, 1912
James Parkinson

THE PATIENT LAY on the operating table with his shaved head clamped to a frame to hold it completely still: a necessary procedure since other parts of his body shook with uncontrollable tremors. He was wide awake and looked extremely nervous as the surgeon started boring into the top of his skull with a drill similar to one used by a dentist. Having made the hole – about the size of a five pence piece – the surgeon pushed a fine probe, three inches long, deep into the patient's brain. When it touched 'the spot', it was the first time the patient's limbs had stopped shaking in eight years.

While the description of this terrifying procedure sounds like it might have been written 200 years ago, the operation was in fact first performed in 1987 and marked a pioneering breakthrough in medical surgery to control Parkinson's disease.[1]

Mike Robins's problems had started ten years previously with a twitch in his right shoulder. Within a few months it had become an uncontrollable tremor down his right side and he was eventually diagnosed with Parkinson's disease. 'I was put on medication, different combinations of tablets, but nothing really worked, and there were a lot of side effects,' Mike said. 'The only alternative to medication, I was told, was surgery, which although it was in its infancy, the results to date had been encouraging, and I was prepared to try anything that might improve my symptoms.' Most patients suffer debilitating side effects after they have been taking medication for about five years. These effects include dyskinesia (violent writhing), hallucinations, psychosis and depression, and they can often be worse than the disease itself. Mike felt that anything was better than the hell he was going through with the drugs and decided to face the operation, even though he would have to remain awake throughout.

Deep brain stimulation, as the operation is called, is now widely used to alleviate a variety of movement disorders, although the majority of operations are carried out in patients with advanced Parkinson's disease. The patient needs to be awake to help guide the placement of the electrode to the area that abates the symptoms of Parkinson's disease, as well as to monitor any untoward effects as the probe moves through the brain. 'When the probe was passing certain parts of my brain, I saw colours brighter and more intense than ever before. When it moved to the area that controls speech I had to keep talking, so that the team would know instantly if anything was wrong,' Mike explained. 'Then the surgeon said there were just a few more millimetres to go, and as soon as he touched the correct spot – an area of the brain about the size of a cashew nut – my right leg and my right arm stopped shaking immediately.'

Once the implant has been placed in the brain, a battery-powered neurostimulator is positioned under the skin near the collarbone; it connects to the implant via a lead that lies under the skin of the scalp. The neurostimulator sends a mild electrical current to the tiny device which delivers electrical stimulation to an area of the brain that controls movement and muscles, known as the subthalamic nucleus. The electrical stimulation modulates the signals that cause many of the symptoms of Parkinson's disease. Mike uses a gadget like a TV remote handset to control his symptoms: when he switches the stimulator off his arms immediately begin to shake so violently that he can hardly get back to the switch to turn it on again. But when he does, the tremors stop instantly. A video of him can be found online; it is very poignant to watch.[2]

Parkinson's disease is one of the most familiar of all neurological disorders and the second most common neurodegenerative condition after Alzheimer's disease. Worldwide, there are about 6 million people with the condition and at least 10,000 new cases are diagnosed each year in Britain alone. More common in the elderly and in men, approximately one in a hundred people over the age of 65 will get Parkinson's disease. As our ageing population lives longer, more and more people will be affected.

It is a terribly debilitating condition. The main symptoms are severe tremor, muscle stiffness and slow movements, and diagnosis is usually based on the presence of two of those three symptoms. There are, however, other common symptoms such as a tendency to fall forward, which leads to a running motion when the patient intends only to walk; problems with communication, such as writing and speech; and a mask-like (expressionless) face due to rigidity. In the advanced stages, swallowing often becomes difficult and walking impossible as the patient becomes more and more rigid. Death is usually due to complications resulting from the symptoms, such as falls and infections.

Mike was talking at a meeting organised by the Parkinson's Disease Society, explaining to journalists and press officers the necessity of medical research on animals. Various drugs, as well as operations such as deep brain stimulation, now provide enormous relief to sufferers of Parkinson's disease, but all these interventions have to be tested on animals first before they can be used on people. Part of my job involved promoting medical research to the media, so I was there to learn how I could support the academics at my institution who did this kind of work in their dealings with journalists. At the time, animal rights activists were very active and intimidating; standing up and advocating research that had been tested on animals risked attracting their attention. But Mike had a powerful argument and as I sat listening to him talk about the operation that had transformed his life, I marvelled at all the research that had gone into developing the technique – it must have taken many decades. However, when the meeting ended I realised there was still one thing I wanted to know – why was it called *Parkinson's* disease? Who was this man the world had forgotten? Who was this Mr Parkinson?

1
Living and bleeding in London

Ah! my poor dear child, the truth is, that in London it is always a sickly season. Nobody is healthy in London, nobody can be.

Jane Austen, 1816
Emma

JAMES PARKINSON was born into the Enlightenment on Friday 11 April 1755. He grew up alongside the Industrial Revolution and died a Romantic on Tuesday 21 December, 1824. His life spanned a period of intellectual turbulence and political upheaval, burgeoning science and technology; they were dramatic and exciting times to be alive. Among his contemporaries were Mozart and Marie Antoinette, the chemists Joseph Priestley and Humphry Davy, the surgeon brothers John and William Hunter, Edward Jenner who discovered the smallpox vaccine, the poet William Wordsworth, the painter William Turner, and the geologist William Smith.[1]

James was the second of six children,* only three of whom – James and two younger sisters, Margaret and Mary – survived

* See the Parkinson family tree on page vi for dates of James Parkinson's siblings and other members of the family.

to adulthood; the three others died within their first five years.[2] There is little known about his mother, besides the fact that she was called Mary and that burial records say she died on 6 April 1811 aged 90 – a grand age for someone of those times. Fortunately, because it was common in the eighteenth century for parents to choose names for their children that honoured their relatives – indeed, to avoid insulting anyone there was even a convention for the order in which the relatives were chosen – it becomes possible to identify that she was a Mary Sedgwick, baptised 16 April 1721 in Rotherhithe, Surrey.[3] As people were generally buried less than a week after their death, it seems likely that Mary actually died a few days short of her 90th birthday.

Attempts to determine the birthdate and birthplace of Parkinson's father, John, have met with less success.[4] Furthermore, there is no record of any marriage between a John Parkinson and a Mary Sedgwick (or indeed any Mary at all) between 1750 and 1753 – the most likely period for a marriage to have taken place, given the arrival of the couple's first child in November 1753.

Nor has any portrait of James Parkinson yet been found, although the internet boasts two different photos supposed to be of him. Unfortunately, since photography was not invented until 1838, fourteen years after Parkinson died, a photograph of him cannot possibly exist. The photographs in question are of two different James Parkinsons. One was a dentist who lived 1815–1895. The image of him was clipped from a group photo taken in 1872 of the membership of the British Dental Association. The man in the other photo, who has a big bushy beard, is a James Cumine Parkinson (1832–1887), an itinerant Irishmen who ended up as a lighthouse keeper off the coast of Tasmania.[5]

We are therefore left to imagine Parkinson's physical appearance, and to help us do that we have a brief verbal description written by his young friend, Gideon Mantell.[6] Mantell would have been in his early twenties when he knew Parkinson, then in his late fifties. He tells us: 'Mr Parkinson was rather below middle stature, with an energetic intellect, and pleasing expression of countenance, and of mild and courteous manner; readily imparting information, either on his favourite science [fossils], or on professional subjects'.[7] Like Parkinson, Mantell was a medical practitioner with a passion for fossils.

Another man with whom 'our' James Parkinson is often confused was an older James Parkinson (1730–1813) whose wife purchased the winning lottery ticket for the disposal of Sir Ashton Lever's exotic natural history collection. Noted for the artefacts it contained from the voyages of Captain Cook, formation of the collection had bankrupted Lever. In order to recover some of the money he obtained an Act of Parliament which allowed him to sell the collection by lottery, but at a guinea each he only sold 8,000 tickets, when he had hoped to sell 36,000. The lucky James Parkinson who acquired the collection spent nearly two decades trying to make a success of Lever's museum, eventually putting it up for auction in 1806.[8] 'Our' James Parkinson was present at the auction and purchased a number of items.

The James Parkinson with whom this book is concerned lived all his life in Hoxton, a village located a mile north of Bishopsgate, one of the narrow medieval gates of the City of London,[9] within the parish of St Leonard's Shoreditch in the county of Middlesex. In a survey of 1735 the total number of houses in Hoxton was 503,[10] but it was fast becoming urbanised as London rapidly expanded northwards during the Industrial Revolution. In 1700, London had a population of just under 600,000; a century later it had reached over a million and was the largest city in the world.[11] Today Hoxton can be found on the enlarged maps which represent the very heart of London in its A–Z of streets. These pages cover an area of less than three miles across from north to south, and five miles east to west, which is larger than the whole of London was in 1750.

As the city became more and more prosperous during the eighteenth century this was reflected in Hoxton where the population grew rapidly. There was a phenomenal rise in trade in the docks and in business generally, which fed an increase in employment and attracted agricultural workers out of the fields and into the metropolis. Residential areas in the city were taken over for business purposes, and as houses were demolished to make room for factories, warehouses and offices, displaced residents and incomers were forced to find homes beyond the City walls in places like Hoxton. The City Fathers, wealthy merchants and businessmen who could afford a horse and carriage, were able to live where they chose and opted for country seats or sophisticated squares. 'Oh how I long to be transported to the dear regions of Grosvenor Square!' sighs Miss Sterling in George Colman's popular comedy *The Clandestine Marriage*.[12] Such Georgian squares launched a new style of town-house: the narrow-fronted terrace; and vertical living became both a novelty and a necessity as space became scarce and land more expensive. Terraces were often set around a square to compensate for the fact that the houses themselves had little land of their own.

The Parkinsons lived at No. 1 Hoxton Square, a three-storey terraced town house constructed between 1683 and 1720 around a large square of more than half an acre.[13] The house was built of bricks since there was a requirement to use fire-resistant materials following the Great Fire of London. In almost every room there were large, open fireplaces carved in an elaborate design. Some of the rooms were connected by elegant arches and many had deep panelling on the walls with pastel colours painted on the ceilings. The most important rooms, impressively large, were on the first floor where long sash windows looked out over the square which formed the focus of this elegant community.

But only the residents were able to enjoy its privileges, each householder owning a key to the garden's delights. From these windows the Parkinsons could see the spire of St Leonard's Church, a fine example of Georgian ecclesiastical architecture.[14] There James was baptised on 29 April 1755, married on 17 May 1781, and buried on 29 December 1824. His grave can no longer

St Leonard's Church, Shoreditch, as James Parkinson would have known it.

be found in the graveyard; it is probably in the crypt along with hundreds of others that were moved there around the beginning of the twentieth century so that the road could be widened.[15] However, a badly deteriorating plaque dedicated to the memory of his father, John Parkinson, can still be seen on the churchyard wall. John had been the much-loved apothecary surgeon in Hoxton for more than 40 years, fulfilling a position in society similar to that of today's GP. The twelve-line inscription probably once told us who had erected the plaque and why, but it is now illegible. Another plaque inside the church, put up in 1955, celebrates the bicentenary of James Parkinson's birth.

The original house at No. 1 Hoxton Square was still standing 100 years ago, although by then it was derelict.[16] At that time it had a smaller two-storey building on the back with a central door that opened on to a side street. This door had probably been the entrance to the apothecary shop where the Parkinsons made up and dispensed medications. Behind that was yet another small building which may have been added at a later date to house Parkinson's ever-growing collection of fossils. Side streets provided access to shops and services, but beyond these, when James was born in 1755, were open fields, market and flower gardens, orchards, and grand old mansions standing in extensive grounds.

Despite the apparent grandeur of Hoxton Square, sanitary conditions were appalling. Household waste fed into open ditches that flowed down the centre of the streets, since Hoxton had no sewer. The ditches discharged into a tributary of the Walbrook river that ran through Shoreditch; the Walbrook, now one of London's several subterranean rivers, eventually released Hoxton's waste into the heavily polluted Thames. At night there would be commodes in the bedrooms, while in most dining rooms there was a set of chamber pots hidden behind curtains or in a cupboard for the relief of gentlemen after dinner. These were generally emptied straight into the street, although one of the greatest causes of pollution of London's waterways occurred with the introduction of the improved water-closet in the 1770s. Many of these overhung streams – the earliest and simplest way of disposing of the contents. The house itself probably had piped water, drawn from the Thames and supplied via pipes of elm wood laid under the main streets, although the source of that water was highly questionable and the contents virtually undrinkable, as one Scottish visitor lamented:

> If I would drink water, I must quaff the mawkish contents of an open aqueduct, exposed to all manner of defilement; or swallow that which comes from the river Thames, impregnated with all the filth of London and Westminster – human excrement is the least offensive part.[17]

Rain water too, 'being, from the soot and dirt on the roofs of houses etc, loaded with impurities' was rarely used, except for the meanest domestic purposes.[18]

Originally the nine water companies in London were each allocated a different region of the City, but when 'healthy competition' was introduced, the result was cut-throat. Each company established separate reservoirs and pumping stations, tore up roads and pavements in order to lay competing sets of pipes, canvassed each other's customers and made wild promises they had no hope of keeping, in order to steal a march on the competition. After a few years of this mayhem, the companies again agreed to divide the City between them and withdrew to their allocated districts, but then a cartel formed which allowed charges to rise steeply in order to pay for the costs recklessly incurred during the previous years of warfare. It's a familiar story.

The primary sources of lighting were candles and oil lanterns, and the only source of heat was invariably an open fireplace burning coal in a cast iron basket, so not the least drawback to living in the City was the constant pall of thick smoke that hung around its shoulders. The travel writer Pierre-Jean Grosley complained that winter in London lasted eight months and that the smoke, 'rolling in a thick, heavy atmosphere, forms a cloud which envelops London like a mantle'.[19] The fallout from this cloud, soot, covered the buildings and anything left outside, even the horses. Aside from the burning of coal in the grates of every household in town, soot was generated by the thousand-and-one small businesses that choked the City's back streets – the smithies, the potteries, the brewing, baking and boiling trades, and the myriad other enterprises. And the pall didn't stay within the City walls: 'the smoke of fossil coals forms an atmosphere, perceivable for many miles', grumbled another tourist;[20] so it can be assumed that even the Parkinsons' fashionable residence, a mile outside the City walls, was covered in grime.

Several good schools existed in Hoxton and the surrounding area and James probably attended one of these, as his published advice on how to prepare young men for the medical profession refers to the need for them to have had a 'common school education'.[21] By the mid-eighteenth century the syllabus of many Middlesex schools included Latin, Greek and French, arithmetic, book-keeping, 'all branches of the mathematics', and the 'use of globes'.[22] Natural philosophy, the precursor to modern science, was introduced on to the curriculum of some private schools around this time, and since Parkinson considered natural philosophy an essential background for medical students, it seems likely he studied the subject at school. In addition to the languages already mentioned, he was able to read German and Italian, and he also used shorthand throughout his life, which he says he learnt as a boy. These were usually subjects for which extra fees had to be paid, suggesting his schooling was of a higher standard than a 'common school education'.

Hoxton, renowned for its 'dissenting' ambience, offered university-level educational facilities in the form of the Nonconformist Hoxton Academy, which moved into Hoxton Square in 1764.[23] Nonconformists advocated religious freedom and opposed State interference in religious matters, but as these beliefs did not 'conform' to the views of the established Anglican Church, nonconformists were restricted from many spheres of public life. They were also barred from various forms of education which compelled them to fund their own academies. The Hoxton Academy provided a university education for young men who were prevented from attending Oxford or Cambridge because of their dissenting views, its presence giving the Square an almost collegiate air. Although James is unlikely to have attended the Academy, since he and his family were members of the Anglican congregation at St Leonard's, the Parkinsons undoubtedly knew many of its tutors, attending them and their students when they were sick. Among the Academy's many well-known pupils was the radical political philosopher William Godwin who studied there for five years between 1773 and 1778.[24] Godwin was just a year younger than Parkinson, so it is possible that they knew each other during this period. Whether or not this was the case, they both acquired a radical social conscience around this time that was to shape the rest of their lives, and they certainly knew each other later on.[25]

For most of Parkinson's life 'Mad' King George III was on the throne. It was a reign dominated by wars: the Seven Years' War (1756–1763), the American War of Independence (1775–1783), and the French Revolutionary Wars (1792–1802), which became the Napoleonic Wars (1803–1815); each was almost twice the cost of the previous one and this hugely increased the national debt. Consequently, everything was heavily taxed, from hair powder to candles, and this burden fell heavily upon the working man. 'More than two thirds of every shilling we earn, is torn from us by Taxes laid on the articles most necessary to the support of life,' complained Parkinson.[26] As a result, the population was frequently in turmoil, riotous and disordered. But such disturbances were a way of life in the eighteenth century – a distraction from the drabness of the life of the poor, a release of emotion and energy – and did little to disrupt the upper classes.

By 1760 London had begun to expand dramatically both north and south of the Thames. To accommodate these developments, London's first bypass was built by two turnpike trusts. It bounded the north side of the metropolis from Paddington to the City, greatly diminishing the congestion caused by the innumerable herds of sheep and cattle wending their way along Oxford Street and Holborn to Smithfield Market. As the Industrial Revolution got under way and such major projects became commonplace, the open fields, gardens, orchards, and fine old houses for which Hoxton was noted fast disappeared.

At the same time, people flocked to cities from the fields at a rate that accommodation could not keep up with. With overcrowding and war came soaring rents. No longer could a family possess its own home but was obliged to share it with others until the little houses became grossly overcrowded. By 1801 the population of Shoreditch was 34,766, crowded into 5,732 houses; the parish later became the most densely populated square mile in the country. What had once been a nice middle-class area was fast becoming populated by the 'lower orders', such that by 1814 the inhabitants of Shoreditch were described as 'Chiefly of the trading community: brewers, dyers, brick-makers, watch and clock-manufacturers, japanners etc, etc. The number of dissenters of all persuasions in this District is immense. The poor are exceedingly numerous.'[27]

There would have been little question that as the eldest son, James was to follow in his father's professional footsteps. So at the age of sixteen, in 1771, he began seven years as an apprentice apothecary to his father, eventually taking over the practice when John died. In turn, James would teach his eldest son, John, who would take over the practice when James died, and John's son James was to do the same when his turn came. Thus at least four generations of John and James Parkinsons worked as apothecary surgeons in Hoxton. It was a family business.

In the eighteenth century there were three types of medical practitioner: physicians, surgeons and apothecaries. Each was overseen by its parent body: the Royal College of Physicians, the Company of Surgeons (which in 1800 became the Royal College of Surgeons of London) and the Worshipful Society of Apothecaries. There was a clearly defined hierarchy, with apothecaries at the bottom of the ladder and physicians at the top. Fellows of the Royal College of Physicians had to have a medical degree from either Oxford or Cambridge and comprised an elite group who supplied healthcare to the rich and were consulted in difficult cases. They were forbidden to practise a 'trade' such as that of apothecary, surgeon or man-midwife.

Below them came the physicians whose religious beliefs barred them from Oxford and Cambridge, and who therefore obtained medical degrees elsewhere. Their qualification licensed them to practise as physicians and to 'dabble in trade'. Surgeons and apothecaries tended to pursue apprenticeships, but the latter remained largely unregulated until the Apothecaries Act of 1815. On completion of their apprenticeship, apothecaries could take an oral exam at the College of Surgeons that qualified them to practise as apothecary surgeons. Many did not – although they still practised surgery.

The apothecary's original role had been to prepare and dispense remedies prescribed by a physician, in the same way that a chemist dispenses medications prescribed by a doctor today, but by Parkinson's time the apothecary would prescribe for minor ailments himself. The majority of town apothecaries, and practically all those in the country, visited patients of the poor and lower middle-class, so would be admitted by the back door to attend the servants of the wealthy, while physicians, arriving in a chaise at the front door, administered to the family. The apothecary would prescribe and supply medicines he had compounded himself, the most common being clysters (enemas) which were recommended for symptoms of constipation and, with more questionable effectiveness, stomach aches and other illnesses.

Apprentice apothecaries were trained to recognise the many plants, berries, roots, barks and minerals used in various remedies, as well as in how to grind and mix the ingredients. On entering the apothecary's shop, customers would have been enveloped in pungent smells, evocative of distant and exotic countries. Large jars of herbs, spices and medications lined the walls, and huge bunches of drying plants hung from the ceiling; drawers contained crushed oyster shells, mercury, dried roots, variously coloured powders, liquids and ointments that were all added to medications, as well as neatly arranged surgical instruments for bleeding and blistering patients or administering enemas; implements on display for compounding and dispensing drugs included a huge pestle and mortar, the apothecary's most important piece of apparatus. Each apothecary had his own jealously guarded recipe book of medications, taken and adapted from published pharmacy books called pharmacopoeias.

When diagnosing a complaint, medical practitioners in the eighteenth century still adhered to the teachings of Hippocrates who, more than 2,000 years previously, had promoted the doctrine of the four 'humours' of the body: blood, phlegm, black bile and yellow bile (or pus). Disease was defined as an imbalance of these humours; thus when an illness occurred it was the physician's role to bring the body's humours back into balance. Some 500 years after Hippocrates this method of treatment had been widely disseminated by Claudius Galenus, better known to us as

Galen, a prominent Roman physician and philosopher of Greek origin, whose theories dominated and influenced Western medical science until well into the nineteenth century. He considered that blood was the dominant humour and the one most in need of being controlled, so he advocated frequent bloodletting, the popularity of which was reinforced after he discovered that veins and arteries were filled with blood, not air, as was commonly believed. In order to restore the balance of the humours, a physician would remove 'excess' blood from the patient or give them an emetic to induce vomiting, a diuretic to encourage urination, or a sudorific to bring on a sweat.

But the tide of knowledge was beginning to turn and in later years Parkinson recalled what a waste of time he felt his seven-year apprenticeship had been. He spent the first five making up medicines, an art he considered could have been obtained in as many months. In the remaining years he learnt bleeding, dressing a blister and, 'for the completion of the climax – exhibiting an enema',[28] the colour and shape of the stool being an important part of diagnosis.

Bloodletting was considered an art. It was a procedure best left to the experienced practitioner, who would know how much to take and where to take it from. 'In ascertaining the quantity of blood to be taken away,' Parkinson explained, 'not only must the sex, age, and strength, be considered; but also the degree of violence of the disease and the importance of the part affected.' In cases where it was necessary to produce an effect on the whole body, blood could be taken from the most 'convenient part', but where a result was required in a specific area, 'bleeding should be employed, as near as convenient to the inflammation'.[29] Bleeding was to be done quickly so that the blood flowed as fast as possible, but one also had to take great care not to take too much for, as Parkinson admitted, there was a danger of 'inducing other diseases, more difficult of removal, than the original complaint'.

On the other hand, if too small a quantity was taken, 'the disease will not be removed'. It was a fine line. Nevertheless, Parkinson advised that everyone should know how to open a vein and draw blood, just in case they had to do it in order to save someone's life when no surgeon was available.[30]

After bloodletting, Parkinson considered that the most powerful method of 'relieving the overloaded vessels and of lessening the disease' was the proper administration of purgative medicines. Purging was a frequent and rapid evacuation of the bowels brought on by administering a purgative such as castor oil or the juice from the skin of an unripe cucumber, but again it was necessary to know exactly what effect this would have for fear of doing more damage than the disease itself. Some purgatives, for example, when used in conjunction with sudorifics to induce a sweat, could be counterproductive and stem the flow of perspiration, 'and thereby occasion an increase of the original complaint'.

Blistering, or cupping, was believed to encourage the flow of blood and clear local 'stagnation'. It was achieved by creating a vacuum in a heated glass cup placed flush against the patient's skin; as the air cooled in the cup, a vacuum formed causing the skin to be sucked up into the cup. When sufficient pus was formed in the blister, it would be opened and the pus allowed to ooze out. Such blisters could be kept open for several weeks in order to obtain the maximum amount of pus. (In fact, by causing localised inflammation, cupping helps trigger an immune system response, so it is possible that in some cases patients did benefit from this treatment.)

Towards the end of his apprenticeship, in February 1776, Parkinson spent six months gaining practical experience at the London Hospital on Whitechapel Road.[31] Depending on the fee paid, students either 'walked' the wards of a hospital, benefiting only by observing the surgeons in action, or, for a higher fee – around £50 a year – they became 'dressers'. Parkinson became a dressing pupil to the surgeon Richard Grindall, which meant he helped with those patients directly under Grindall's care, assisting the surgeon in dressing wounds, mending fractures, performing some of the lesser operations of surgery and, of course, bleeding patients when necessary. Each surgeon could be responsible for up to six dressers at any one time.

Grindall, who had been appointed to the surgical staff more than 20 years previously, was known for his operative skill and for his dedication to his profession. 'He was also a great Oddity,' recalled one student, 'but a perfect Gentleman in his appearance and manner, never seen . . . but in a well-powdered wig, silk stockings and shoe buckles.'[32] Often appearing in the wards late at night, Grindall would check on the progress of his patients who were critically ill. This attention to those in his care was unusual as surgeons did not stay on the premises and night visits were normally left to the hospital's resident apothecary, or the dresser on duty.

Surgeons and physicians would work in the hospitals free of charge, earning an income by taking dressing pupils and apprentices, by giving lectures and by developing a private practice which they would often run alongside their hospital work. But establishing a practice could take many years, so it was necessary to first build up a good reputation in the hospital, since this would later attract the paying public to the practice.[33] Connection with a great hospital was therefore extremely important. As a result, hospital posts were highly sought after and could be contested as fiercely and as expensively as seats in a parliamentary election. Unfortunately, in all the seven 'great' London hospitals at this time, there were only 22 posts for physicians and 23 for surgeons, so opportunities were few; an aspiring physician or surgeon had to wait until someone either resigned or died before a post became vacant.[34] One such surgeon waiting in the wings at this time was William Blizard.[35]

After serving a surgical apprenticeship in Surrey, Blizard came to the London Hospital to study under the surgeon Henry Thompson, whom he eventually succeeded when Thompson died in 1780. While waiting for a position to arise, Blizard joined the London's Board of Governors, a position he held at the time Parkinson was studying there. Although a man of extremely high standards who demanded a great deal from everyone around him, Blizard also had the ability to inspire; his one-time student John Abernethy recalled, 'I cannot tell you how splendid and brilliant he made it appear'.[36] Twelve years older than Parkinson, Blizard initially mentored the younger man, but as they came to realise they had similar political and religious views, and shared an interest in fossils, they developed a close friendship that would continue throughout their lives.

The sick and injured who flocked to the London Hospital came largely from the wharves of Rotherhithe and the workshops of Spitalfields, Aldgate, Wapping, Whitechapel and West Ham. Among the common medical ailments Parkinson helped Grindall treat were pneumonia, tuberculosis and infectious fevers. Many of the workers – women and children, as well as men – also suffered from scalds and burns, lacerations, fractures and crush injuries sustained during their work. Burns and scalds were common injuries in the home as well, with women in particular suffering terrible accidents when their highly combustible clothing caught fire. Should this happen, Parkinson advised first calling for help without opening the door – as the external air rushing in would immediately increase the progress of the flames – and then to 'tear off that part of the clothing which is in flames'. If in a parlour, the burning woman was told to seize the water jug and pour it over herself. For this reason alone, the jug should be large and always kept full. If that did not put out the flames, she was to sit down on the floor, since standing was more likely to 'render the communication of the flames to the upper part of her dress', and smother the flames using the hearth carpet which, Parkinson recommended, should always form part of the furniture in every room, in case of such an emergency.[37] The Parkinson establishment was undoubtedly a model household, equipped to cope with all such eventualities.

When Parkinson was studying at the London Hospital, apprentices, pupils and dressers were expected to attend lectures in various subjects likely to improve their medical knowledge. But although hospitals such as Guy's and St Thomas's had lecturers based there, the London, being on the edge of town, did not offer much in the way of these additional facilities. Only lectures on surgery and physic (the practice of medicine) were available, but if a student wanted to gain any semblance of a complete medical education, he would also need to take courses in anatomy, midwifery, medicine, natural philosophy and chemistry. All of these were on offer elsewhere in the City, but a student 'must

An artist's impression of the new London Hospital, Whitechapel, in 1752.

necessarily neglect part of his business at the hospital', traipsing around London on foot in search of them.[38]

It was entirely up to the student to choose which course to study and as each course charged a separate fee – around £3 to £4 – the less well-off might decide not to attend any at all. Furthermore, all dressers were required to be on hospital premises between 9am and 2pm, and again in the evenings between 6 and 9pm, so there was little time available for study. Years later Parkinson wrote angrily about the system for educating medical students, and what he considered to have been the 'misdirection' of his studies. Being placed behind the apothecary's counter for seven years and receiving his hospital education in the manner described was 'absurd', making him feel he had been 'robbed of his fair chance of becoming proficient in his profession'[39] – for it was during his time at the London Hospital that Parkinson decided to become a surgeon.

Towards the end of the eighteenth century, surgery had become a glamorous profession, many considering it had 'arrived at such a degree of perfection in Great Britain as leaves no room for France any longer to boast of her supremacy. A journey to Paris is no longer necessary to complete a Surgeon's education.'[40] This success had been brought about by John Hunter, Henry Cline and others whose speed and skill with the knife had made them celebrities. The rich and famous flocked to be treated by them and many a young surgeon aspired to similar heights, despite having to 'submit to long and heavy – and even wearisome plodding, through the paths of science'.[41] As a young man Parkinson was no exception, although later in life he warned that celebrity could not be 'cheaply purchased', for the student had much to learn before achieving the 'fame for which he pants'.[42] As well as the study of natural philosophy, physiology, chemistry and physics, along with French and German in order to read 'the numerous scientific works which are written in the

French and German tongue', an aspiring surgeon needed, above all, a good understanding of anatomy. It was the 'key-stone' to becoming truly proficient, and the only way to achieve the level of skill demonstrated by John Hunter was through the practice of dissecting dead bodies.

The Company of Surgeons was first established in 1745 and since that time criminals executed on the gallows had provided cadavers on which young men training to become surgeons could test their dissecting skills. A large number of eighteenth-century statutes specified death as the penalty for minor property offences and anyone stealing goods worth more than five shillings could be sentenced to hang.[43] Execution was a public spectacle meant to act as a deterrent to crime and huge crowds would assemble to watch as the prisoners were transported through the streets. On arrival, they were stood in a horse-drawn cart, blindfolded and their hands tied together. The noose was then placed around their neck and the cart pulled away, leaving the condemned to hang until they died, which often took several minutes. Friends and relations might pull on their legs to help them out of their misery. After the execution, unseemly struggles for possession of the corpse often broke out between assistants to the surgeons who needed the body for teaching purposes and friends of the prisoner who wanted to give the victim a proper burial.

As demand among the surgeons had increased, the price of a cadaver had risen to over £100, and by 1750 grave robbing had become an alternative source of supply.[44] But after the introduction of the Murder Act in 1752 it was no longer legal to give an executed murderer a proper burial, so instances of fighting over the corpses became less common. After that the bodies were taken straight to the Company of Surgeons for dissection and, as the surgeons received a regular supply of cadavers, the price fell dramatically. In the years immediately following the Murder Act, costs dropped as low as £3 8s, later stabilising at around £12 13s.

The Company of Surgeons had opened its newly built operating theatre, Surgeons' Hall, in 1753, implementing a system of electing Masters, Wardens and Stewards of Anatomy to administer it. The Masters lectured in anatomy, the Wardens demonstrated dissection during lectures and made sure that

'The body of a MURDERER exposed in the Theatre of the Surgeon's Hall', *Newgate Calendar*, 1794.

everything was conducted 'with Decency and Order', and the Stewards dissected and prepared the bodies for the lecturer.[45] Unfortunately, such was the unpopularity of these posts, which were 'voluntary' and considered a burden on people who were already extremely busy, that heavy fines had to be imposed on individuals not accepting their 'election'. Even then many refused, preferring to pay the fines, which became an important source of income for the Company. But because of this difficulty in obtaining lecturers and wardens, over the ensuing 20 years the Company of Surgeons had become moribund and its reputation as a training establishment had declined. At the same time, several well-known surgeons had set up their own private schools of anatomy, considered by many to be more modern in their teaching methods and superior in their premises and facilities.

The year before James started training at the London Hospital, 1775, his father had been elected as Warden at Surgeons' Hall, a position he had chosen to accept, rather than pay the fine, and which he held for two years. So when James began his training, he attended the anatomy lectures at Surgeons' Hall, where he watched his father demonstrate the dissections. The cadaver would be cut up by the Steward before the lecture and only the parts to be discussed that day were taken into the theatre, since a whole cadaver would rapidly decay due to the body heat emanating from the students. The Warden, having demonstrated the elements they were to observe, would then pass round each body-part, instructing students to examine it in detail before handing it on.[46] Meanwhile, the Master would explain its purpose, drawing important elements on the blackboard in order to illustrate how the parts interacted with each other. Parkinson, ever critical of prevailing teaching practices, considered this an inappropriate way in which to demonstrate anatomy. He argued that chopping up the body beforehand and presenting it to the student in many separate pieces over a long period of time made it difficult for them to acquire an overview of the whole 'human fabric' and to understand how all the parts were related to each other.[47] Perhaps because of such criticisms, later that year a committee was set up to review the Company's method of training students. As a result, the two unpaid Master of Anatomy positions were abolished, their place being taken by a Professor of Anatomy on a salary of £120 a year.

Parkinson evidently studied hard, doing the best he could in difficult circumstances, but being a student in the big city seems to have changed little over the past 200 years, and however diligent he tried to be, 'public spectacles, feasts, balls and assemblies daily hold out their temptations; and pleasure, under every fascinating form, will seek to secure you in her snares'. Friends were a distraction too: 'Your friend invites you to accompany him to the play but you, knowing that you have some evening lecture to attend, beg to be excused. Your objection is opposed by the observation "you really make too much of a slave of yourself – you must have some little amusement" and so you are enticed to go out.' Having given way to such inducements, the evening cannot be concluded 'without a bit of supper and a glass of wine. The glass circulates freely, until the conviviality of the evening renders your attendance at any lectures impossible.' Next morning, 'Your head aches and your spirits are low, you therefore trust to a friend for his notes of the evening lecture, but these are not so intelligible as your own. Thus three or four lectures are lost to you.'[48] Generally, though, Parkinson was a model pupil almost desperate to take advantage of his time at the London. Even so, having only six months in which to complete his studies meant he finished his training as a dresser still feeling 'miserably ignorant'.

2
The hanged man

It is with the utmost satisfaction I can inform you of a case in which I have been able to restore to life one, who before the institution of your Society, would probably have been numbered with the dead.

John Parkinson, 1777
Transactions of the Royal Humane Society

AT SEVEN O'CLOCK in the evening, on Tuesday 28 October 1777, there was a loud rap at the door of No. 1 Hoxton Square. John Parkinson was urgently summoned to the house of Bryan Maxey who, as the messenger informed him, had hanged himself. John called for James to accompany him and they immediately set out for Maxey's house about a quarter of a mile away. There they found Maxey who had apparently been dead for half an hour: a coldness had already spread over his body and his jaw had become so fixed that they had to use considerable force to move it. A woman in the house, aware that a reward was offered by the Humane Society to those who resuscitated the dead, had tried with some small success to keep Maxey warm by rubbing his stomach with a flannel. In addition, a neighbour who practised bleeding had taken about eight ounces of blood from Maxey's arm before the Parkinsons arrived.

Father and son immediately began to give Maxey mouth-to-mouth resuscitation, a new and controversial technique introduced by the Humane Society just a few years previously. James, now 22 and in the final year of his apprenticeship, performed most of the physically taxing work, since John was in poor health. Eventually, as John reported to the Humane Society a few days later, they were successful in returning Maxey to consciousness. After 40 minutes of resuscitation he had given a deep sigh which was followed by an increase in the strength of his pulse. After an hour he was able to breathe without assistance and in another half an hour he regained consciousness. 'He then complained of an excessive pain in the head,' explained John, but once given some warm brandy and water, followed by a 'purging' broth which enabled him to produce a stool, Maxey was deemed well enough for the Parkinsons to leave.[1] Producing a stool was, of course, a vital sign of life.

By the end of a week, Bryan Maxey was completely recovered except for a slight dimness of vision and a feeling of numbness on the right side of his head. He expressed the utmost sorrow for his 'crime', as suicide then was, and gratitude to those who were instrumental in restoring him to his wife and children. This satisfactory result was honoured by the Humane Society with the award of its Silver Medal to the young James Parkinson who had worked hard to restore Maxey to life. It became one of his most treasured possessions and he proudly left it to his eldest son in his will.

We are not told why Maxey tried to kill himself, but he appears not to have been alone in the quest to find an end to his misery. According to the Reverend Caleb Fleming,[2] a dissenting minister and neighbour of the Parkinsons, there was an epidemic of 'self murder' around this time – a period of political and financial instability caused by the American War of Independence.[3] As the Reverend Fleming explained, the war had resulted in 'an alarming shake to public credit', as well as an 'obstruction to trade and commerce' which in turn caused widespread unemployment. The price of food rose sharply and while the rich indulged themselves 'in every debauchery and extravagance' imaginable, the poor starved. Unable to cope, many committed suicide: 'The insolvent and dissatisfied are cruelly laying violent hands on themselves in great numbers,' lamented Fleming.[4]

Had Maxey died, the punishment for his crime would have been the forfeiture of all his goods and chattels, as well as those belonging to his wife. So the family would have lost not only a husband and father, but all its worldly goods as well. Even the Reverend Fleming, who considered the heinous crime of self-murder to be an act of High Treason against the sovereignty of the Lord, felt that punishing those left behind was too severe. Instead, he proposed 'the naked body [of the suicide] should be exposed in some public place', over which the coroner would deliver an oration on his terrible crime. The body, like that of the murderer, should then be given to the surgeons who would use the parts to demonstrate anatomy to their students. Such a dreadful punishment would have terrified the likes of Maxey, since the idea of dissection after death contravened a belief in the sanctity of the grave where the dead body was supposed to rest undisturbed until Judgement Day when it would be reunited with the soul. If the body was dissected, it would never find its soul, which would wander around in Purgatory forever. Maxey was grateful Parkinson saved him from this fate worse than death.

The Humane Society had been founded just three years earlier by two doctors who were concerned by the number of people wrongly taken for dead and subsequently buried, or dissected, while still alive. The *Newgate Calendar*, a popular book that reported the crimes, trials and punishments of notorious criminals, quoted the words of a surgeon about to dissect a murderer recently taken down from the gallows:

> I am pretty certain, gentlemen, from the warmth of the subject and the flexibility of the limbs, that by a proper degree of attention and care the vital heat would return, and life in consequence take place. But when it is considered what a rascal we should again have among us, that he was hanged for so cruel a murder, and that, should we restore him to life, he would probably kill somebody else. I say, gentlemen, all these things considered, it is my opinion that we had better proceed in the dissection.[5]

The Humane Society advocated artificial respiration and recommended warming the body, administering stimulants, and bleeding, provided the latter was done with caution. For some years, blowing tobacco smoke into the rectum (fumigation) was also considered beneficial, bellows being used 'so as to defend the mouth of the assistant'.[6] The Society stressed the importance of prompt and prolonged treatment – no case should be abandoned unless vigorous efforts, maintained for at least two hours, had been unsuccessful. Parkinson, however, recommended attempting resuscitation for no less than three or four hours, disdainfully considering it 'an absurd and vulgar opinion, to suppose persons irrecoverable, because life does not soon make its appearance'.[7]

Keen to promote these new resuscitation techniques, the Humane Society initially offered money to those rescuing someone from the brink of death. A handsome reward of two guineas was distributed among the first four people to attempt the rescue, and four guineas paid if the person survived. But a scam soon became widespread among the down-and-outs of London: one would pretend to be dead while the other brought him back to life, and they would then share the reward money between them. Consequently, monetary rewards were soon replaced by medals and certificates.

A network of 'receiving houses' was set up in and around the Westminster area of London where bedraggled bodies, most of them pulled out of London's waterways, could be taken for treatment.[8] But rare cases of apparent lifelessness also occurred when people were struck by lightning. Ten years after the Bryan Maxey incident, a house in Crabtree Road near Shoreditch Church was badly damaged by lightning and two men, one passing by and one in the house, were struck and seriously injured. The injured passer-by was brought into the damaged house and appeared to be dead, although by the time James Parkinson arrived fifteen minutes later a faint pulse was perceptible. His hands and legs resembled those of a corpse, being excessively cold and of a dark, almost black, colour; his head was bent right back and despite strenuous efforts to bring it forward, it was immovable; his eyes were red, each eye staring in a different direction, which gave him a wild appearance. A large red streak like lightning had appeared down his right side, along with several lesser ones on his legs, the skin in those places being badly scorched. When the man was revived sufficiently to speak, he complained of pain in the head and chest, the latter aggravated by a frequent cough which threw up a lot of blood. His fellow sufferer, the man in the house, had similar symptoms but was not spitting blood. One of the metal buttons on his sleeve had melted, as had the buckle on one shoe where his foot was badly burnt. A red streak, about two inches wide, was evident all down his right side, right arm and on the shins of both legs, forking in a manner similar to lightning and with what appeared to be small 'sparks' coming off it; from all these lesions he felt a considerable burning pain.

Parkinson considered the symptoms shown by these men suggested 'a congestion of blood in the head and lungs' so he promptly removed six ounces of blood from each patient. They were then put to bed, their burnt legs and hands wrapped in flannels wetted with an ointment composed of sweet oil and

The ward of the Receiving House of the Royal Humane Society in Hyde Park. An electrical machine stands on the table between the two windows.

ammonia, and a draught was given them for neutralising any acid in the stomach; this was 'washed down with half a pint of weak brandy and water, as hot as could be drunk'. The two men immediately broke into a sweat and fell into a deep sleep from which they awoke a few hours later apparently recovered, although the man in the house still felt a burning pain where the lightning had 'so beautifully marked him'. The passer-by, when he happened to meet Parkinson a few weeks later, informed him that although he was by then fully recovered, a pain in his hands and legs had come back so severely the day after the incident that he had been confined to bed for a fortnight.

So intrigued was Parkinson by the effects of lightning on the 'animal system' that he wrote a paper about the incident – his first publication[9] – which has subsequently been recognised as a benchmark for the description of lightning injuries.[10] What struck Parkinson most was the ability of lightning to contract the muscles, notably in the man whose head had been thrown backwards in an immovable spasm, and which only relaxed after he had slept for a couple of hours. In a similar case in which another patient of Parkinson's had been struck by lightning across his face, the man had apparently gone blind. On examining him Parkinson concluded that the muscles surrounding his eyes had gone into spasm, forcing his eyelids to shut so tightly that the man had been completely unable to open them for seven months. With a great deal of effort, Parkinson prised them apart and within a couple of days the blind man was able to see again quite normally.

Interest in electricity had been stimulated by Benjamin Franklin's experiments in the 1750s that proved lightning was electricity.[11] Twenty years later, the potential for using electricity for 'revivifying' lifeless patients was just being introduced. Parkinson, always quick to pick up on modern developments, attempted to apply the technique to a young man who had drowned. The man had been in the water for over an hour and others had unsuccessfully tried to resuscitate him for over two hours before Parkinson arrived, whereupon, as he reported, 'having in my pocket a portable electrical machine, I was induced to make a trial of the effects of electricity'. He first tested the strength of the 'electrical fluid' by giving himself a minor shock while holding the dead man's arm to complete the circuit. To his amazement, the extended arm, hand and fingers of the corpse convulsively 'bent at every joint', despite rigor mortis having already set in. He then went on to perform an early form of defibrillation by giving the man electric shocks above his heart, as recommended by the Humane Society, but unfortunately the experiment did not bring the young man back to life. It was interest in such experiments, and the ability of electricity to apparently raise people from the dead, that encouraged one reviewer of Mary Shelley's *Frankenstein* to declare that the novel had 'an air of reality attached to it by being connected with the favourite projects and passions of the times'.[12]

Over the following years, knowledge of the Humane Society's methods for saving lives spread throughout the country and proved so successful that by 1794, twenty years after the Society was founded, some 2,582 cases of lifelessness had been reported, of which 1,835 persons had been restored to life (an astonishing 71 per cent).[13] In 1800, as part of the country's centenary celebrations, the Lord Mayor of London, Vice Presidents of the Humane Society, and 350 'prominent gentlemen' assembled on the Society's anniversary to watch a 'glorious procession of restored men, women and children' march past, accompanied by appropriately solemn music.[14] Perhaps Bryan Maxey was among them – he would have been 52.

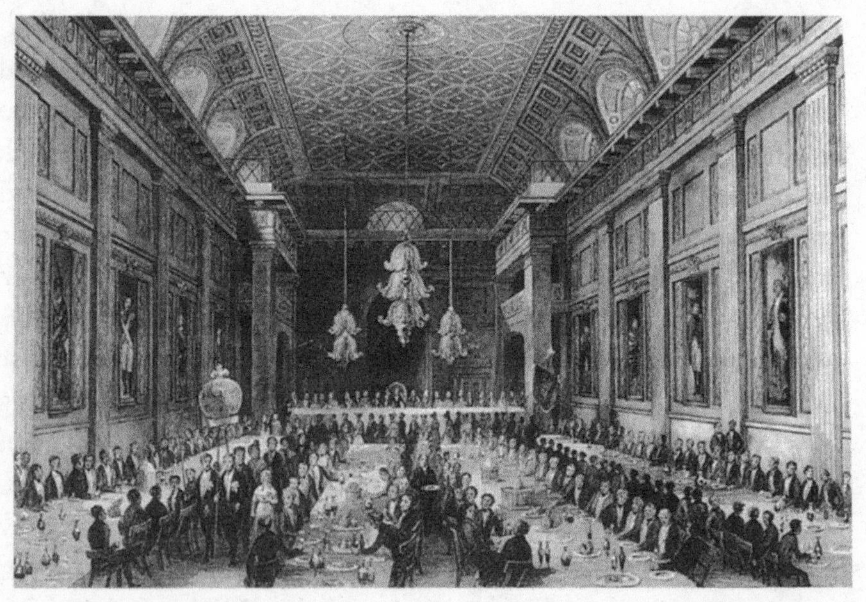

A Royal Humane Society dinner at the Freemasons' Hall, London. The diners watch a procession of persons saved during the year.

3
Fear of the knife

I began a scream that lasted unintermittingly during the whole time of the incision and I almost marvel that it rings not in my ears still! so excruciating was the agony.

Fanny Burney, 1812
Frances Burney: journals and letters

THREE YEARS after completing his apprenticeship to his father, James married his childhood sweetheart, Mary Dale, on Saturday 12 May, 1781. Mary also lived in Hoxton, in Charles Square, which was just a hundred yards or so from Hoxton Square. Mary's father, John Dale, was a silk merchant but came from a medical family – his eldest brother and his father were both apothecaries in Hoxton, and the middle brother had graduated as a physician from the University of Leiden in the Netherlands. Mary's great uncle was the celebrated apothecary, botanist and geologist, Samuel Dale.[1] Thus the Parkinsons and the Dales mixed in the same circles, had similar interests, and the several generations of medical men would have known each other well. James and Mary probably grew up together and may have been expected to marry from an early age, although by the middle of the eighteenth century, couples were beginning to have more choice in their selection of spouses and relationships were generally formed on the basis of personal affection, rather than obeying parental wishes.

Samuel Johnson's dry comment on courtship is as true today as it was then: 'A youth or a maiden meeting by chance, or brought together by artifice, exchange dances, reciprocate civilities, go home and dream of one another . . . find themselves uneasy when they are apart, and therefore conclude that they shall be happy together'.[2] But marriage meant subordination for women: 'By marriage, the husband and wife are one person in law: that is, the very being, or legal existence of the woman is suspended during the marriage, or at least is incorporated and consolidated into that of the husband: under whose wing, protection and cover she performs every thing,' pronounced Sir William Blackstone when describing the woman's place in 1777.[3] Furthermore, married women would spend much of their day with female kin and neighbours, while men spent their time with other men. The couple would often eat apart, walk apart, even, most of the time, sleep apart, and there is no reason to think that James and Mary's marriage was different. After their wedding, the couple settled down to family life with James's parents in the large house in Hoxton Square, so at least the two Marys, James's wife and his mother, had each other for company when the men were out. It also meant that James was still on hand to help his father in the practice.

Since the age of 40 John Parkinson had suffered badly from gout and it was getting progressively worse; now, for two or three months of the year he was incapacitated by the pain and swelling of his joints, and James was required to do most of the work in the practice. James recalled how, during a particularly bad attack, his father was visited by the eminent physician Dr Hugh Smith, whose public lectures the Parkinsons had attended because, as the advert said, he was to address 'new thoughts on the nervous system, the gout and paralytic complaints'. He was evidently an expert on the subject. Smith prescribed Blackrie's Lixivium for John's gout, since a 'gentleman of respectability' that he had known, who had taken it for treating gravel (kidney stones), found that it had also cured his gout. The recipe for this drastically strong alkali concoction was quite simple:

> Take eight ounces of pot-ash and four ounces of quick-lime fresh from the kiln; mix and put into a glazed earthen vessel; then pour upon them a quart of boiling soft spring-water; let the infusion remain twenty-four hours, stirring it now and then; and afterwards filtrate it for use.[4]

Blackrie particularly recommended the quicklime be made from oyster shells. The dosage was from '30 drops twice a day up to two teaspoonfuls thrice a day, in weak veal broth or any other soft mucilaginous vehicle'.

Having paid a substantial sum for Dr Smith's attendance, John Parkinson was inclined to take his advice, although James had concerns that the continued use of such stringent medicines might be harmful. Nevertheless, John took the Lixivium for over a year and at the same time studiously avoided wine and other liquor. During this period he was completely free from gout and so at the end of the year decided he was well enough to resume his previous habit of drinking 'stale porter', a malt beer that was kept longer than usual in order to improve its flavour. But the gout returned and late in 1783 John experienced an acute pain beneath his pubic bone which was relieved only by large and frequent doses of opium. Eventually he developed ascites, an accumulation of fluid in the peritoneal cavity which results in the belly swelling to an enormous size.[5] Although most commonly due to advanced liver disease, ascites is also a symptom of many types of cancer. James thought the problem was due to bowel cancer and since there was no effective treatment he could only try to ease his father's suffering with opium. John Parkinson died on 10 January 1784, aged 59, and one cannot help wondering if Blackrie's Lixivium was a contributing factor.

Despite his ambitions to become a surgeon while studying at the London Hospital, James had so far not taken the Surgeon's Diploma which would formally entitle him to practise as one. His father had obtained his diploma at the age of 40, so with one qualified member in the family, perhaps it was felt that there was no need for James to go to the expense of taking the exam as well. Instead, he had joined his father as a partner in the business and continued to gain surgical experience by performing minor operations under his guidance. But when John died, James, still only 28, was left to operate the practice single-handed. Perhaps conscious of how easy it would be for patients to drift away if they felt they were not getting the same standard of service as his father had provided, James took the examination for the 'great diploma'[6] at Surgeons' Hall just three months after the funeral. Considered by many to be something of a farce, the assessment included such basic questions as: 'What is a compound fracture?' and 'Pray, what are ye. muscles of ye. Abdomn?'[7] There was absolutely no requirement to demonstrate a practical knowledge of anatomy, or that one had any skill at all with the surgeon's knife. The oral assessment lasted about fifteen minutes, after which a fee of fifteen shillings was demanded.[8] Nevertheless, it was imperative to gain the diploma as it entitled the holder to become a member of the Company of Surgeons. Accordingly, James Parkinson's name appeared on a list of surgeons approved by the London Medical College on 1 April 1784, just ten days before his 29th birthday.

Although now a formally qualified surgeon, Parkinson continued to update his medical knowledge and in the winter of 1785–86 he enrolled on a six-month course of lectures on the Principles of Surgery given by the legendary surgeon and anatomist John Hunter.[9] Hunter was then the most celebrated surgeon in London. He was at the height of his popularity, treating many famous personalities, and at the peak of his intellectual powers, contributing paper after paper to the Royal Society.[10] Hunter's one-hour evening lectures were held at his house in Leicester Square every Monday, Wednesday and Friday evening between October and April – about 80 lectures in total. Parkinson attended diligently, although there must have been times when it was difficult for him to travel the three miles from Hoxton, particularly during inclement weather which frequently made the roads impassable. Although the Thames did not freeze over as it had the previous two winters, the severe frost and snow in early January 1786, the heavy rains in mid-January, and the deep snow at the end of the month would have turned the streets of London into a quagmire, making it almost impossible to get around.[11] Nevertheless, the notes James took suggest he did not miss a single class.

Hunter's newly refurbished premises had only just been completed and Parkinson was among the first group of students to be taught there. When he arrived for the inaugural session at seven o'clock on 10 October 1785, he would have found a spectacular brick and glass three-storey construction built between Hunter's fashionable residence at 28 Leicester Square and the building at the back that fronted on to Castle Street. On the ground floor was a huge reception room lined with paintings by contemporary artists that led to a spacious operating theatre where Hunter lectured and demonstrated his dissections. His students would be seated on raked wooden benches arranged in a semicircle so that everyone could get a good view. Occupying the other two storeys of the building was his purpose-built museum in which Hunter housed his collection of anatomical and natural history specimens.

As a schoolboy, Parkinson had learnt the 'art' of shorthand, which he now found invaluable since he was able to take down Hunter's lectures verbatim – he wasn't going to miss a single word. The great benefit of taking notes in this way, he argued, was the 'unremitting attention which they necessarily excite . . .

preserving the mind from straying, so that nothing material can escape unrecorded'. Parkinson would transcribe his notes either when there was a 'hiatus' during lectures or when he got home afterwards.[12] Such was Hunter's legacy that in 1833, almost fifty years after the lectures had been given, James Parkinson's son John, by then a surgeon himself, considered it worth publishing his father's transcriptions of Hunter's lectures under the title *Hunterian Reminiscences*. He did this, as he explained, in order to give current students an understanding of the giant on whose shoulders they were standing, for 'it is pleasing ... occasionally to descend from the height to which modern surgery has attained, and carefully retrace each step, until we arrive at the very base on which it rests, every stone of which may be said to be inscribed with the name of JOHN HUNTER'.[13] In the Preface to this work, John eulogises Hunter, comparing him to the great Isaac Newton in his search for 'sublime truths', and declaring that he was 'gifted with a penetrating genius'. However, John also admits that 'there exist[s] in many obscure passages ... the acknowledged difficulty that Mr. Hunter had in always making himself understood'. Like many brilliant men, it seems John Hunter was not a very good lecturer. Indeed, on reading *Hunterian Reminiscences*, one must sympathise with James Parkinson who originally had to make sense of it all. The lectures covered a myriad of subjects, few of which we would consider 'surgery' today, and Hunter flitted from one topic to another in rapid succession which made them difficult to follow, as John Parkinson rather harshly pointed out:

> The Editor is very far from considering these 'Reminiscences' as constituting a system of surgery; they must rather be regarded as consisting of fragments (and those often very unconnected), but containing most of Mr. Hunter's original pathological doctrines.[14]

Nevertheless, some of Hunter's ideas undoubtedly influenced James Parkinson's thinking. When discussing gout, for example, which James's father had experienced so acutely and which he himself would suffer from in years to come, Hunter questioned whether the condition was entirely hereditary or could also be brought on by overindulgence:

> The gout also has been supposed hereditary; but we find the gout attacking some whose parents have, perhaps, from all eternity, been free from the gout; as is the case where luxuries are introduced where they have not yet been known, or where the native of such a place is removed to where luxuries are abundant.[15]

However, Hunter was mistaken in thinking we also inherit a predisposition for smallpox, illustrating how poorly understood this disease was at the time:

> And, on the other hand, there are diseases equally hereditary, which have never been so esteemed, because the hereditary disposition has been overlooked . . . thus it must be allowed that the smallpox is as hereditary as gout and scrofula, we being born with a susceptibility of its particular action . . . and hence many families have it worse than others.

Hunter briefly mentions the palsy, the condition which would make Parkinson so famous, when describing the case of a woman who probably had what we now call Parkinson's disease:

> A lady, at the age of seventy-one, had universal palsy: every part of the body shook which was not fully supported. The muscles of respiration were so affected, that respiration was with difficulty effected; but in sleep the vibratory motions of the muscles ceased, and the respiration was performed more

equably: any endeavour of the will to alter these morbid actions increased them.[16]

This description may have stayed in Parkinson's mind, for years later he would point out that in the true shaking palsy, the affected limb *continued* to shake even when supported. Hunter, however, was correct in stating that all tremors ceased when the patient was asleep, a point Parkinson was mistaken about when he came to writing his own account of the disease.

Whatever we think about Hunter's lecturing ability, Parkinson evidently gained a great deal from these lectures as he praised Hunter's abilities whenever the opportunity arose; no doubt they also stimulated his determination to become a first-rate surgeon.

Once Parkinson had gained his Surgeon's Diploma he would have regularly carried out minor operations – all without the use of anaesthetics. The chemist Humphry Davy would suggest in 1800 that nitrous oxide could be used to relieve pain, but it would not be routinely employed until 1840. But despite the fact that a high percentage of patients died on the operating table or shortly afterwards due to shock from the pain or loss of blood, Parkinson had no compunction in promoting the value of surgery, berating women in particular for their fear of the knife in the removal of breast cancer tumours. He explained how the tumour became more difficult to extract the larger it got, and strongly recommended its early removal since, during the early stages of the disease, 'the operation is comparatively trivial, and requires but very little time for its performance; the pain which accompanies it is very far short of that which is imagined'. Having the operation, he assured women, would almost certainly exempt them from a recurrence of the disease which would otherwise kill them.

Recognising that in many cases the fear of the operation – 'the assemblage of the surgeons, the preparation of instruments, and many other circumstances' – was much worse than the operation itself, he offered this somewhat perplexing advice on how patients should prepare themselves psychologically:

> The mode of thinking which should therefore be adopted, by those to whom such an operation has become necessary, is to let the mind dwell only on the absolute pain of the operation, abstracted from all foreign circumstances and visionary terrors . . .[17]

When the English novelist Fanny Burney discovered that she had breast cancer in 1811, she took the momentous decision to have her breast removed.[18] Her description of the operation confirms Parkinson's observations regarding the terror patients felt as preparations were made; her description of the pain, however, suggests that it exceeded even her worst nightmares. Having waited several weeks to hear when the operation would be performed, Fanny was suddenly given two hours' notice in which to prepare for her ordeal. As was common at the time, she was to have the operation at home:

> I strolled to the Sallon – I saw it fitted with preparations, and I recoiled . . . the sight of the immense quantity of bandages, compresses, sponges, Lint – made me a little sick: – I walked backwards and forwards till I quietened all emotion, and when became by degrees, nearly stupid – torpid, without sentiment or consciousness; – and thus I remained till the Clock struck three.

When the time came, with tremendous fortitude Fanny lay down on the bed that had been prepared in her salon and a veil was spread over her face; she closed her eyes and resigned herself to endure 'a terror that surpasses all description':

> . . . when the dreadful steel was plunged into the breast – cutting through veins – arteries – flesh – nerves – I needed no injunctions not to restrain my cries. I began a scream that lasted unintermittingly during the whole time of the incision and I almost marvel that it rings not in my ears still! so excruciating was the agony.[19]

Fanny went on to describe every terrible detail of the operation, including how she could feel the knife 'rackling against the breast bone – scraping it!' The operation to remove her breast lasted twenty minutes and for many months afterwards she was in deep shock and could not think of her ordeal without feeling sick, but slowly she recovered. Fanny Burney lived until she was 87, a fine example of someone who overcame her fear of the knife and who proved, as Parkinson predicted, that the early removal of a breast tumour could save a woman's life.

In the years following his father's death, James worked tirelessly, his day starting around 6.30am when he opened up the apothecary shop and not finishing until at least twelve hours later. Although now qualified as a surgeon, his main income still derived from his gruelling work as an apothecary. The arduous daily toil was described by another of his profession:

> Of all the branches of the medical profession, that of the apothecary, without doubt, is of most consequence to the health of the nation at large. In this city [London] where a physician attends one patient, an apothecary attends twenty; and, in the country, this proportion is more than doubled. "He is," says a celebrated writer, "the physician to the poor at all times, and to the rich whenever the disease is without danger." . . . Huts, hovels, and cottages which . . . inclose such infinite numbers of human beings, that feed, with perpetual pabulum, diseases of the most infectious and fatal tendency, compose almost

> exclusively the walk of the apothecary. To him is likewise allotted the care of nearly all prisons and poor-houses.[20]

In Georgian times medical practitioners typically visited their patients, rather than have them call in to the shop or practice, and Parkinson probably did his rounds on foot, covering many miles every day. He complained that patients all too frequently requested he attend them right at the end of the day, regardless of 'tempestuous weather', and that they expected him to respond immediately he received their summons. Furthermore, if the message arrived after he had set off on his rounds, it would be necessary to go out again, 'giving him the trouble of pacing over a mile or two of ground, which he has already trodden'. As a consequence, he warned, these patients might not always receive the best treatment because the apothecary was so distracted, worrying about all the other things he should have been be doing while visiting them so late. Officially apothecaries were only supposed to charge for the cost of the drugs they prescribed, but as the role of the apothecary turned into that of the general practitioner, as it did during Parkinson's lifetime, the cost of drugs went up in order to pay for his time as well. The average cost of a visit was around one guinea, although the poor would have paid a lot less, and often Parkinson would accept payment in kind, particularly produce for his table.[21]

Highly conscientious, Parkinson was acutely aware of how the lives of his patients depended on his decisions and what might happen if things went wrong. Plaisters* and ointments frequently contained mercury, which did little to cure the patient and much to poison him with afflictions at least as unpleasant as those from which they were suffering. Bleeding – generally performed with an unsterile instrument – could become

* Pastes applied to the skin for healing or cosmetic purposes.

dangerously septic.[22] When assessing such risks, Parkinson knew that not only was his professional reputation at stake, but he was also liable to accusations of impropriety for the measures he had adopted. If he cured a patient it was more by luck than design, but should they get worse or even die, he would be to blame. Scepticism – even cynicism – towards physicians and apothecaries was as old as the profession itself. The New Testament told physicians to heal themselves. That, according to Benjamin Franklin, was the last thing they could do: 'God heals and the Doctors take the Fee,' he famously pronounced. The learned and pious Dr Ridgeley was of the opinion that 'If the world knew the villainy and knavery (beside ignorance) of the physicians and apothecaries, the people would throw stones at 'em as they walked in the streets',[23] and even the great physician Thomas Sydenham believed that many poor people owed their lives to their inability to afford treatment. Parkinson thought this attitude unfair, feeling that his time and labour were often ill-used, but although he was aware that sometimes the cure caused more problems than the disease, medicine was insufficiently advanced for him to identify what he was doing wrong.

During the last two decades of the century, the organisation of medical practice was going through a fundamental change, as the role of the apothecary transmogrified into that of the general practitioner. The three-way division into apothecaries, surgeons and physicians was gradually replaced by our more familiar two-tier hierarchy of general practitioner and consultant. This change was largely driven by the fierce competition between these professions for fee-paying patients, none of whom wanted to pay more than necessary, so both rich and poor would go to whichever practitioner they felt could help them best for the least outlay. In theory, patients would call in a physician to diagnose their complaint, employ a surgeon to let blood, and have an apothecary make up the medicines. In practice, only the rich could afford to consult them all, so the neat divisions were rarely observed and physicians, surgeons and apothecaries aggressively competed with one another, each complaining that those on the rung below were stealing their patients. Apothecaries, at the bottom of the ladder, blamed the quacks who made outrageous claims for their counterfeit medications.

But while the apothecary was encroaching upon the domain of the surgeons and physicians, the pharmacist was taking over the dispensing activities of the apothecary, and even beginning to prescribe over the counter. The role of the pharmacist (or, more strictly, chemist and druggist)[24] had evolved alongside that of the apothecary, whose seven-year apprenticeship was unaffordable to many. As Parkinson has already told us, learning the 'art and mystery' of manufacturing medications did not really require such a long-term commitment. As a consequence, many druggists and chemists also produced medicines, although they did not diagnose and prescribe. As apothecaries moved into a more advisory role, pharmacists further developed their own area of expertise, which put them into direct competition with the apothecaries who were still involved in the preparation and supply of medicines. Recriminations between these two groups appeared in a pamphlet war published during the second half of the eighteenth century, in which each blamed the other for serious improprieties. The apothecaries accused the pharmacists of selling and using impure foreign drugs, refuse, dross, and adulterated articles in compounding prescriptions, as well as leaving out all the more costly ingredients. The pharmacists accused the apothecaries of taking 'monstrous profits', and their incompetence, illiterate character and dishonest practices were portrayed in graphic detail.[25] It must have been difficult for apothecaries of Parkinson's high standing to read such provocative material.

Gradually, each discipline widened its field of expertise, the boundary between apothecaries and surgeons blurring the most, until the general practitioner emerged, able to deal with most types of medical complaint but calling in a consultant for difficult cases. By 1800 William Charles Wells, a physician at St Thomas's Hospital, felt confident enough to remark on 'the complete establishment of the apothecaries as medical practitioners'.[26] But none of this made Parkinson's job any easier – if anything, it just added to his load.

By the time James and Mary celebrated their tenth wedding anniversary, in May 1791, they had considerable family responsibilities. Although only four of their seven children were to survive into adulthood, by the end of 1791 they were the proud parents of five children: eight-year-old James John, six-year-old John William Keys (named after James's close friend and brother-in-law to be, John Keys), four-year-old Emma Rook, two-year-old Jane Dale, and baby Henry Williams – Wakelin and Mary Dale were still to come. Then, tragically and unexpectedly, on 21 January 1792, two-year-old Jane died. Although James would have known from his practice that child mortality was high, her death must have been devastating and it seems he never quite forgave himself for not being able to save her. His later references to 'the high degree of guilt' felt by those left behind, 'the tormenting reflections, which must harass the minds of those who become convinced of having sacrificed their friend or relation', and the 'afflicted, self-accusing survivors' who may have unwittingly caused a death, are indicative of the terrible grief and torment experienced by the couple on the death of their child. Perhaps it is no coincidence that within weeks of this tragic event, James made a momentous decision that was to change his life.

4
The radical Mr. Parkinson

If anyone asks me what a free Government is,
I answer . . . it is what the People think.

Old Hubert, 1793
Pearls Cast Before Swine

THE 1790S, often called 'the age of revolution', were unsettled times. After decades of labouring under the burden of massive taxes due to a succession of expensive wars, tensions were high amid the working classes, from whom there emerged a persistent and vociferous call for political reform and the working man's right to vote. Despite William Pitt the Younger being elected Prime Minister in 1783 on a promise of social reform, the common man (and woman) still did not have a vote. Before 1831, only male owners of freeholds who paid more than 40 shillings (£2) a year land tax were eligible to vote. This small number represented about 2 per cent of the population – which left the vast majority disenfranchised. Furthermore, boundaries between constituencies had existed for more than 300 years, never being adjusted for changes in the rapidly shifting population, so the huge industrial towns recently created by the Industrial Revolution had virtually no parliamentary voice. Manchester, with a population approaching 250,000, had no MP at all, whereas the 'rotten borough' of

Gatton in Surrey, with its population of 40 people, of whom only seven could vote, had two MPs.

Over in France the heavily downtrodden populace had also become extremely restless and agitated, such that in July 1789 over 1,000 Parisians stormed the Bastille prison, this outbreak of violence effectively marking the start of the French Revolution. An abortive rebellion in June 1792 was followed by a successful one on 10 August when an armed mob stormed the Tuileries Palace where King Louis XVI and his family were being held under surveillance. The King was imprisoned and a Legislative Assembly established which decreed that a 'National Convention' should draw up a new constitution. On 21 September the Convention held its first meeting. It immediately abolished the monarchy, set up the Republic and proceeded to try the King for treason. His execution was carried out by guillotine on 21 January, 1793.

Initially, enthusiastic support for the French Revolution among Britain's hungry and severely deprived working classes grew rapidly. There was a long list of complaints against the Government that caused the people to be sympathetic to their revolutionary comrades in France: they were subject to laws made by a government that the vast majority had no say in electing; workers were not paid a living wage; heavy taxes had been imposed to finance a war they did not support, and men could be carried off to fight in that war without notice or compensation. At the same time, men of influence appointed their friends to positions of power and gave them and themselves generous lifelong pensions, paid for from the public purse. With no formal way of registering their discontent, a restless people began to form political societies in order to lobby for the working man's right to vote. They embraced political objectives drawn directly from France, wanting to replace royal and aristocratic rule with representative government.

As these events unfolded, Parkinson's political views began to crystallise as he mixed with radical sympathisers such as the political lecturer John Thelwall and the publisher Daniel Eaton, both of whom supported the French Revolution.[1] Parkinson had first met Thelwall in 1787, probably at meetings of the Physical Society, one of London's first medical societies, and for several years they were close friends.[2] Weekly meetings of this society were held in the operating theatre at Guy's Hospital between October and May, and were open to physicians, surgeons, apothecaries and their pupils. Throughout his life Parkinson tried to keep up with the latest advances in medical science, so he regularly attended in order to hear a dissertation, exchange medical news and discuss cases of interest:

> Mr Bureau acquainted the society he had been informed that the deadly nightshade was a certain cure for madness . . . He promised to make further enquiries respecting this medicine and to give the society all the information he could get on this subject.
>
> Mr Haslam reported the case of a child of three who menstruated – her breasts were considerably enlarged but she had no desire, he believed, of coition.
>
> Mr Home presented an account of a child with a double head. 'It is a species of *lusus naturae* [a whim of nature] so unaccountable that, although the facts are sufficiently established by the testimonies of the most respectable witnesses, I should still be diffident in bringing them before you were I not enabled at the same time to produce the double skull itself.'[3]

Such reports were a constant source of fascination.

John Thelwall was not a medical man but, like many of his contemporaries, he had interests in a wide range of subjects. In 1787 his *Poems on Various Subjects* appeared to some praise and later on he became a close friend of the poet Samuel Taylor

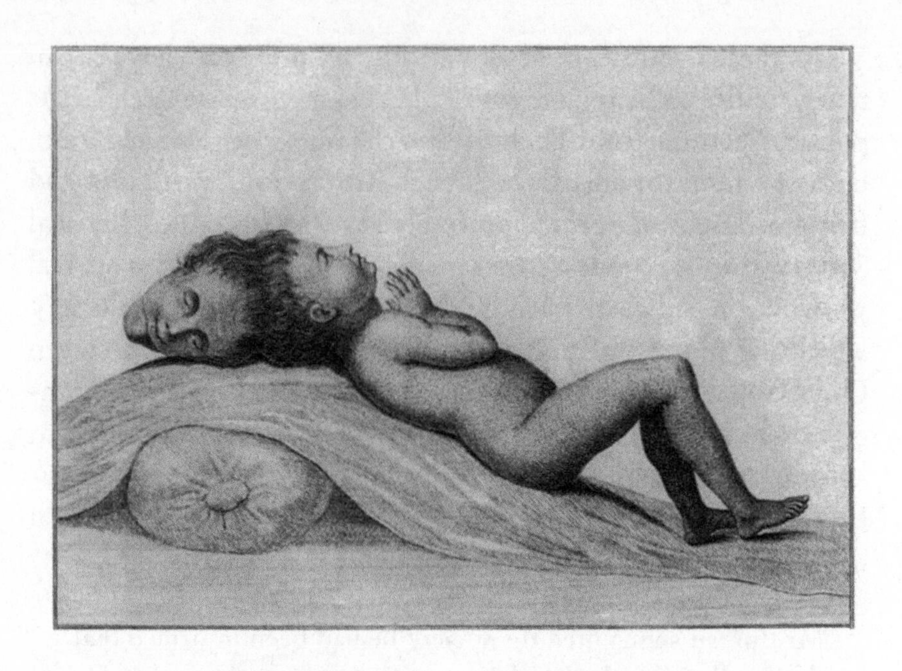

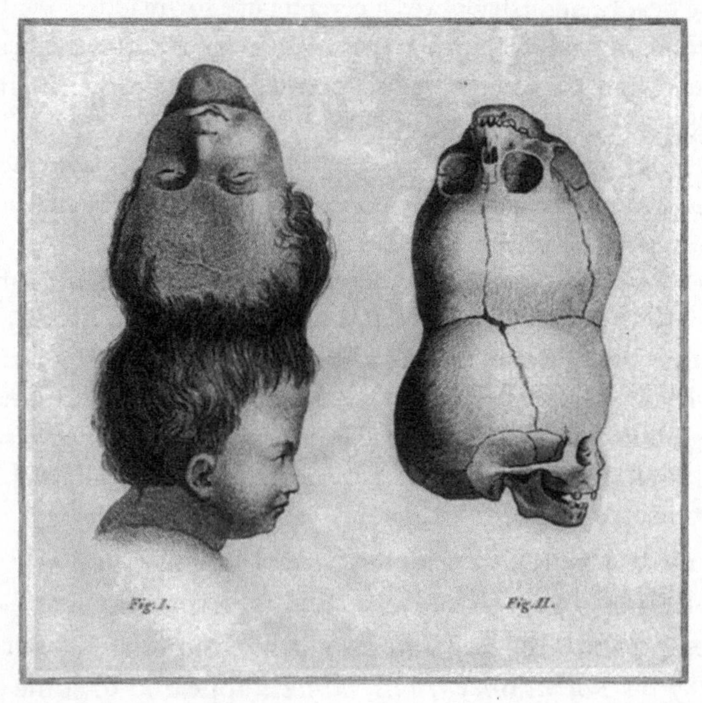

Sketches of the child with two heads, and its skull.

Coleridge. He also attended the anatomical lectures given by Henry Cline who probably introduced him to the Physical Society.[4] There Thelwall gave a dissertation one evening entitled 'An Essay Towards a Definition of Animal Vitality',[5] which was a great success and excited such keen interest that discussion continued on 'six successive nights to a theatre particularly thronged', as he boasted in a footnote when the essay was published.

The two surgeons Henry Cline and Astley Cooper were joint lecturers in surgery and anatomy at St Thomas's Hospital and as a consequence were prominent members of the Physical Society. Cline was much in favour of the French Revolution, had influence with leading men in Paris and had long been a friend of John Horne Tooke, a controversial politician and principal figure in the British radical movement who had been in the public eye for more than twenty years. Thelwall was also a follower of Horne Tooke, whom he regarded as his 'intellectual and political father'. Astley Cooper too, perhaps influenced in his political beliefs by Cline who had been his teacher, held radical views and was to spend time in revolutionary Paris in 1792. There he witnessed the atrocities of the Terror, writing later, 'A revolution may sometimes be a good thing for posterity, but never for the existing generation for the change is always too sudden and violent'.[6] Thus during most weeks throughout the season this close-knit group of radical sympathisers had the opportunity to discuss not just medical cases, but also the increasingly desperate events occurring in France.

Parkinson's sympathy and support for the French Revolution and what it was trying to achieve increased throughout the early 1790s, as his social conscience and his ideas on equality of wealth were honed, and as his frustration over the lack of social reform in Britain grew. He sought out and associated with other like-minded men, officially becoming a member of the reform movement when he joined the Society for Constitutional Information (SCI) around 1790, attending their meetings whenever he could escape from his practice. He was probably encouraged to become a member by his friends in the Physical Society, for at that time the SCI was led by John Horne Tooke whom they all revered. But the SCI drew its members from the upper echelons of society, taking pride in its exclusivity, and working men were not generally accepted into its ranks, a condition Parkinson was not comfortable with. He did not remain a member for long.

Next Parkinson became a Freemason, joining The Lodge of Freedom and Ease in June 1791. Having moved from the Strand the previous year, The Lodge met at the Three Jolly Butchers in Hoxton Market Place, just a short distance from Parkinson's house.[7] In theory the Freemasons' constitution barred political discussion, but during the 1780s Freemasonry became involved in the reform movement and some lodges were hotbeds of radicalism. However, by the 1790s a transformation was occurring and the brotherhood which for seven decades had tolerated a wide range of political views began insisting that 'the true principles of masonry inculcate an affectionate loyalty to the King and a dutiful subordination to the State'.[8] Such expectations would have been an anathema to Parkinson for by that time he had become deeply involved in the radical movement,[9] which was probably why he stopped attending meetings after March 1793.

Still searching for a society which resonated with his own values, Parkinson eventually joined the London Corresponding Society (LCS), shortly after its inauguration. From the start the LCS advanced an entirely new form of politics: ideas and discussion were to be the property of all; the weekly subscription would be low and political organisation was no longer to be confined to the elite or the educated. Its objective was simple:

to obtain parliamentary representation for all men. But such a scheme was a dangerous and radical concept in 1790s England, so when Thomas Hardy, a London shoemaker and founder of the LCS, approached a friend with his draft manifesto, the response was: 'Hardy, the Government will hang you'.[10] Nevertheless, on 27 January 1792 Hardy and eight others met at the Bell Tavern in Exeter Street, off the Strand, where they decided to form the LCS. Hardy was elected Secretary and Treasurer, and Maurice Margarot, a merchant of High Street, Marylebone, and a 'gentleman of elegant manners, a scholar and marked for a leader', was elected Chairman.[11]

The LCS held its inaugural meeting just six days after the death of Parkinson's daughter Jane; within a short time he had joined its ranks, whereupon he rapidly became a prominent and active member. From its inception the LCS played a key role in leading the campaign for political reform. Hardy later estimated, perhaps optimistically, that within six months of its foundation the Society had a membership of 20,000, a number that exceeded those currently eligible to vote. Most of the members were, like Hardy and Margarot, craftsmen or owners of small businesses. Hardy was a strict disciplinarian and although most meetings were held in public houses, no smoking or drinking was permitted. No matters tending to cause disunion or disloyalty to the crown were allowed, and there would be no religious disputes. 'Man,' the LCS manifesto proclaimed, 'may justly claim Liberty as his birthright', and as a member of society it became an 'indispensable duty to preserve inviolate that Liberty for the benefit of his fellow citizens'.[12]

The means by which the London Corresponding Society communicated its ideas across the country was, as its name suggests, by written pamphlets. Sold for a penny or so, thousands and thousands of these pamphlets were published, stirring the collective consciousness and acting as a clarion call to action. But as the reform movement grew, the Government became nervous as it looked across the English Channel at developments in France – could similar events occur in England? Would the rabble rise up, kill the aristocracy and behead the King? Already publications were being sold that the Government believed incited this kind of activity. In particular, Thomas Paine's hugely popular book *Rights of Man*, which advocated political rights for all men and condemned hereditary government, was seen as a serious threat to the country's stability. In it Paine had outlined a plan for a 'Convention' to review English law and suggested that only those found 'worthy' should be retained to participate in this new Government. Paine's message – that he was advocating reform along the lines taken by the French, who had just formed a new Convention – could not have been more evident.

In an attempt to suppress such publications, on 21 May 1792 a Royal Proclamation was issued against all such 'wicked and seditious writings'. This was specifically aimed at halting the publication of cheap editions of Paine's book which the masses could afford to buy, but it also meant that anyone writing or publishing anything that criticised the Government was liable for prosecution. Large posters of the Royal Proclamation went up all over London, calling on magistrates to search out the authors, publishers and distributors of such works and punish them accordingly. To facilitate these directives, magistrates set up a network of spies who were instructed to infiltrate societies such as the LCS and report on their activities. But despite the Proclamation – or perhaps because of it – James Parkinson, together with his friend Daniel Eaton, decided that now was the time to enter the dangerous arena of writing and publishing radical pamphlets.

Daniel Eaton[13] had moved to Hoxton in 1786 to open a stationery shop and it was probably then that he first met Parkinson. Eaton and Parkinson were of a similar age, shared similar backgrounds as respectable members of the 'working trades',[14] had children of a similar age, and held similar political beliefs. Their friendship had grown over the following six years such that when, in the autumn of 1792, Eaton was offered an opportunity to open a new shop closer to the City,[15] the move also marked a turning point in Parkinson's life. Eaton acquired a printing press and embarked on a new venture as printer, publisher and bookseller, and as soon as he was ready to start publishing, it was Parkinson who provided the material. Using the pseudonym of 'Old Hubert', *The Budget of the People, collected by Old Hubert* became Parkinson's first radical pamphlet and Eaton's first radical publication.

Published in late 1792,[16] the pamphlet was a collection of political sketches that satirised and harangued a Government widely perceived to be incompetent and corrupt:

> Among the enemies of the human race, foremost appear the petty PRINCES OF GERMANY [the King and his son], who enact the ravages of despotism on a smaller, and consequently more oppressive scale; who plunder the peasant to maintain absurdly disproportionate establishments; who drag him from his home, the son from his parents, the husband from his family, to form under the rigour of military discipline the instruments of new exaction; who fell the blood of their subjects to swell the pride of a master, and have the insolence to call this GOVERNMENT.

Liberally quoting from philosophers such as John Locke, its bold and outspoken disapproval of the Government and the nation's leaders, its criticism of the war with France which took men from their families and had resulted in a high national debt and excessive tax burden, and its demands for proportional representation of the people in Parliament became Parkinson's and

Eaton's hallmark. Five other pamphlets written and published by them followed in quick succession during the first six months of 1793.

Parkinson enjoyed writing and had a talent for it, so when the LCS formed a Committee of Correspondence in September 1793 in order to communicate its message to other societies and organisations, Parkinson took on the task of drafting and revising many of the pamphlets the Society issued to publicise its views, which Eaton then published.[17] Parkinson later recalled that it was at this stage of his life that he developed the habit of staying up late at night in order to fit in all this extra work. With the day's medical work done and the family asleep in bed, he could concentrate on his political commitments. Sitting in his candlelit study with a warm fire glowing in the hearth, he would write into the early hours of the morning, despite knowing he would have to get up after only a few hours' sleep to visit his patients.

Almost as soon as they began publishing radical pamphlets, Eaton attracted the attention of the authorities. He was warned to cease selling such 'seditious' works and admonished by the Lord Mayor of London himself, but Eaton took no notice and in June 1793 he was tried at the Old Bailey for publishing the second part of Paine's *Rights of Man*. A government spy had been sent to Eaton's shop on two occasions to purchase copies of this work so that they could be used as evidence against him. Accused of 'unlawfully, wickedly, maliciously and seditiously' publishing 'a certain scandalous, malicious and seditious Libel', he faced several years' imprisonment if found guilty.

Eaton was tried before the Recorder of London, the most senior judge of the Central Criminal Court, but he was successfully defended by an able young barrister. Felix Vaughan was a member of the LCS and a passionate believer in every man's right to vote; he was to make his name defending at such trials.

The jury returned after just two hours, finding Eaton 'guilty of publishing, but not with a Criminal Intention'. Vaughan interpreted this unusual verdict as 'not guilty', but the prosecuting counsel vigorously disagreed. The judge, after some deliberation asked the jury to reconsider, but the foreman was adamant: 'We have given our verdict and we persist by it.' Unable to rule on whether publishing without criminal intention constituted a libel or not, the judge decided that Eaton must be committed to prison until the next session, when the point would be further examined. Infuriated, Vaughan demanded that Eaton should be let out on bail while waiting for this decision, but the men who had previously stood bail for him had already left the court. Despite others offering to vouch for him, the judge would only accept bail from the original gentlemen and Eaton was forced to spend the night in Newgate Prison before being allowed out the following day. In the event, the charges were never formally dropped or pressed, because only a month later, Eaton was on trial again.

This time the 'seditious' pamphlet published by Eaton was another of Paine's publications, *Letter Addressed to the Addressers, on the Late Proclamation*, in which Paine had responded in stinging terms to the Royal Proclamation forbidding publication of such pamphlets, as well as replying to the charge of seditious libel made against him for writing *The Rights of Man*. The prosecution had reason to believe that Eaton had entered into correspondence with 'sundry Persons in foreign parts' for 'criminal and wicked purposes'. This clearly alluded to Eaton's communications with Paine who, by this time, had fled to France for fear of imprisonment. While Paine was exiled in Paris, his *Letter* was printed in London on 16 October 1792 by the printers Henry Delahoy Symonds and Thomas Clio Rickman, to be sold at a price of 4d each or a guinea per hundred. When both printers were prosecuted for publishing Paine's

Letter – Rickman escaping to join Paine in France and Symonds being jailed for two years – Eaton stepped in to continue its publication, presumably communicating with Paine in France in order to do so. Eaton did not put his name to the publication as Symonds and Rickman had done, stating instead that it was printed 'for the booksellers'. Nevertheless, it was not difficult for the authorities to discover who the printer was.

On this occasion Eaton was tried before a hand-picked Special Jury, the members of which were selected to try cases of unusual complexity or importance. The judge was the highly unpopular Lord Kenyon, the Lord Chief Justice, about whom anecdotes abounded regarding his lack of a university education, his ignorance of Latin, his coarseness and his bad temper. He was also well known for supporting the Government's view on seditious libel.[18] Eaton was again defended by Felix Vaughan who quoted liberally from Paine's *Letter*, demonstrating the reasonableness of Paine's arguments and even exclaiming at one point, 'Good God! Is everything with the name of Thomas Paine [on it] to be libellous and criminal?' Once again, the jury was persuaded by his arguments and, familiar with the verdict brought in at Eaton's previous trial, they too found Eaton 'Guilty of publishing' but they did not find him guilty of libel.

Kenyon, furious with this outcome from his hand-picked jury, bellowed, 'The verdict should be *Guilty*, or *Not Guilty.*' Not in the least intimidated, the juryman replied, 'We have considered the case, my lord, maturely, and this is the only verdict we can give.' Kenyon responded, 'I do not know the meaning of the words *Guilty of Publishing* . . . The country will expect a [proper] verdict from you.' Whereupon the jury withdrew again and in about 40 minutes returned with a verdict of 'Guilty of publishing that book'. Defeated, Kenyon gave in and Eaton was again allowed to go free, thanks, as Eaton acknowledged, to the common sense of twelve good men.[19]

Spurred on by his success, Eaton began to collect material for a new endeavour and on 21 September 1793, he published the first issue of *Hog's Wash*, a fortnightly publication designed to spread political enlightenment among the masses. Parkinson, still writing under the pseudonym of 'Old Hubert', became one of Eaton's chief contributors and, with his assistance, *Hog's Wash* prospered.

Hog's Wash, and multiple other references to 'swine', 'hogs' and 'pigs' had become the trademark of the reform movement. It was a reaction to a phrase in a work by Edmund Burke, an Irish Whig politician, often regarded today as the 'father' of modern conservatism. *Reflections on the Revolution in France*, written by Burke in 1790, was an attack on the French Revolution and the radicals in Britain who sought to bring about similar changes. Having decried the French Revolution and extolled the virtues of good manners and religion, Burke outlined his fears that should such a revolution occur in Britain, education and learning would be 'cast into the mire and trodden down under the hoofs of a swinish multitude'.[20]

This reference to the masses as 'a swinish multitude' excited public indignation to fever pitch. For years afterwards the term was bandied about in every form imaginable. The first issue of *Hog's Wash* contained a typically sardonic ditty by Parkinson on the front page:

> Since Times are bad, and solid food is rare,
> The Swinish herd should learn to live on Air:
> Acorns and Pease, alas! no more abound,
> A feast of Words, is in the HOG TROUGH found.
>
> *Old Hubert*

In a similar vein, Parkinson also wrote *An Address, to the Hon. Edmund Burke. From the Swinish Multitude* as well as *Pearls*

Cast Before Swine, by Edmund Burke. Scraped Together by Old Hubert. There was a penny magazine called *Pig's Meat; or, Lessons for the Swinish Multitude*; J. Sharpe published *A Rhapsody, to E***** B**** Esq. Ornamented with a humorous print of "The Swinish Multitude"*; and many, many others signed their letters and pamphlets 'a pig', 'a hog', 'a brother grunter' or 'one of the swinish multitude'. A particular favourite was a song entitled 'Burke's Address to the "Swinish" Multitude!' by Thomas Spence, sometimes hailed as England's 'first modern socialist', that was sung to the tune of 'Derry Down, Down, Down'.[21] The first verse went as follows:

> Ye base swinish herd, in the sty of taxation,
> What would ye be after, disturbing the Nation?
> Give over your grunting – be off to your Stye,
> Nor dare to look out if a KING passes by.
> Get ye down, down, down – keep ye down!

J. Sharpe's humorous print of 'The swinish multitude', circa 1793.

Hog's Wash ran for 60 numbers between September 1793 and March 1795, although after a couple of issues it was renamed *Politics for the People*. Partly original material, partly excerpts from other works, and enlivened by satire, irony, humour and verse, the magazine was hugely successful. Throughout the series ran an irregular, but always 'to be continued' 'Sketch of the Most Memorable Events in the History of England from the Landing of Julius Caesar, to the Reign of William the Conqueror'. Written by Parkinson masquerading as Old Hubert, it was a clever parody of the Government, using examples from history to illustrate its current failings. The opening sketch relates Britain's invasion by Julius Caesar and gives us a glimpse of Parkinson's ironic sense of humour:

> Not thoroughly convince'd of the blessings which flow from being *Governed by a Foreigner* [a reference to George III being German] . . . many efforts were made by the Britons to regain their independence, the one most deserving of going on record being that of the illustrious Boadicea.

Boadicea's husband, Prasutagus, King of the Iceni tribe, had hoped to obtain peace with the invaders by bequeathing half of his kingdom to the Roman Emperor, 'for Kings had already begun to consider their subjects as transferable property,' comments Old Hubert dryly. But the Roman procurator, 'modestly conceiving this legacy not to be sufficiently ample for the Emperor his master, took possession of the whole'.

> Boadicea ventured to remonstrate, but the august representative of Caesar, considering that such unpardonable presumption demanded exemplary punishment, ordered the widow Queen to be scourged in the manner of a roman slave, and took on himself the trouble of violating the chastity of her daughters.

> Boadicea, not being sufficiently civilized to bear tamely such injuries at the hands of the Tyrant, placed herself, with her daughters at the head of 20,000 Men, and gave battle to the Romans.[22]

Throughout the pages of *Politics for the People*, Eaton and Parkinson consistently advocated votes for all, annual parliaments, peace among nations, education of the poor, and unfettered discussion of politics and religion. They complained constantly about the way the King and his ministers levied heavy taxes on the poor and appointed large pensions to themselves while the people were starving and their young men were sent off to a war they did not support. But in the eighth issue Eaton overstepped the mark, publishing two items that appeared to directly threaten the monarchy.[23] One of these was a story by 'Citizen' Thelwall, which blatantly compared the King to a cruel, tyrannical cock who dominated the farmyard from the top of a dunghill. Eventually the narrator gets so fed up with this animal strutting about the farmyard and showing off his fine feathers that he decides the only thing to do is to 'rid the world of tyrants' by cutting off its head. With the execution of Louis XVI by guillotine having occurred just a few months earlier, the analogy and the implied threat to the King were unmistakeable.

Two weeks after publication, on 4 December 1793, Eaton was for the third time indicted on a charge of seditious libel. Despite the fact that he had diligently appeared at his two previous trials, the judge, in the hope of bankrupting Eaton, required him to furnish bail of £1,000 for himself, plus two sureties of £500 each. Unable to find such an enormous amount of money, Eaton was imprisoned in Newgate until his trial on 24 February 1794. Notwithstanding this, the regular publication of *Politics for the People* was resumed from 14 December, with Eaton directing operations from his cell and Eaton's wife and son, with help from Parkinson, putting them into practice on the outside.

Eaton was again to be defended by a young lawyer, John Gurney,[24] who was sympathetic to the cause. Initially things seemed to go against Eaton at the trial, but Gurney argued that the article had been an indictment of tyranny in general and feigned dismay that anyone could possibly think that Eaton had intended the cock to be a caricature of the King. Gurney even went so far as to suggest that it was the Attorney General himself who was guilty of seditious libel for daring to propose that Eaton had represented George III as a tyrant.[25] Everyone laughed uproariously. As the jury retired to consider their verdict, they requested a copy of the article to read for themselves, but whether this was because they wished to examine its contents is debatable; it may simply have been for their amusement. Either way, within an hour they returned with a clear verdict of 'Not Guilty'. Eaton was again acquitted and the prosecution

Daniel Isaac Eaton

ridiculed for thinking that Eaton's remarks about the cock referred to 'our said lord the King'. Evidently the jury was made up of reformist sympathisers for no one could have mistaken the real meaning of the article.

Such was the relief at the verdict, the London Corresponding Society had silver medals struck commemorating Eaton's acquittal. Eaton himself, having recently moved his business once more to new premises, triumphantly adopted the imprint 'Printed by D.I. Eaton at the Cock and Hog-Trough, No. 74, Newgate Street'. Eaton's shop quickly became a gathering place for reformers and radicals. In a later issue of *Politics for the People*, Citizen Parkinson wrote a poem to his friend, celebrating his acquittal but also issuing a word of warning for him to be more careful in future. The poem is a rather tedious seventeen verses long, but its essence is encapsulated by the following:

> TO DANIEL ISAAC EATON, on the verdict obtained from an honest jury, on an indictment for libel, tried in February, 1794, at the Old Bailey.

> So, *Dan*, once more you have escap'd
> From Newgate's *noxious* air;
> Th' indictment, surely, was not shap'd
> With all a lawyer's care.
> . . .
> No thanks to him for getting free
> Are due from you, my bold one!
> For he'd be pleas'd were he to see
> You tortur'd by the *Old One*.
>
> No; tis' to Gurney you're in debt,
> And to an honest Jury:
> Vile hirelings now may fume and fret,
> Since you've escap'd their fury.

But here my verse I must not end,
Until I've crammed your ear full
With sage advice, and, as a friend,
Taught you to be more careful.

OLD HUBERT

Parkinson may not have been much of a poet, but he was obviously concerned for the safety of his friend. Despite his unease, he himself continued to work tirelessly for the cause, writing political articles on behalf of the LCS and meeting privately with fellow members. It was not unusual for them to knock on his door at midnight if there was something urgent to discuss.

These were demanding times for the London Corresponding Society as prosecutions increased and sentences became more severe. In August 1793 a young Scottish lawyer, Thomas Muir, was found guilty of sedition for having read out an address from the United Irishmen – another radical group – which complimented the Scottish nation for its spirit of reform, and for having distributed copies of Paine's *Rights of Man*. Muir was prosecuted by Robert Dundas, the Lord Advocate, who was known for handing out severe sentences to reformers, before five judges and a jury whose members had been selected because they all belonged to a loyalist club. Muir's friends urged him to employ an eminent defence lawyer but Muir unwisely decided to defend himself. Though he did so with panache, he was unanimously found guilty and given the unexpectedly harsh punishment of transportation to Australia for fourteen years. A month later, the Reverend Thomas Palmer, a middle-aged Unitarian clergyman of Dundee, was sentenced to transportation for seven years. Palmer's 'crime' was that he had been responsible for the printing and distribution of a pamphlet proclaiming the need for the reform of parliamentary elections.

The severity of the sentences imposed on Muir and Palmer marked a new phase in the Government's repression of reform societies. Earlier convictions for seditious activities had usually brought a sentence of no more than two years' imprisonment, transportation normally being reserved for thieves. Reform societies throughout Britain were shocked and scared. The Scottish societies, which previously had had little to do with their counterparts in England, now appealed for their support. They asked that English delegates be sent to a convention due to be held in Edinburgh in November 1793. Most of the English organisations declined the invitation, but the London Corresponding Society accepted. Accordingly, on 24 October, a large, but orderly, crowd of about 4,000 people attended a mass meeting in Shoreditch to select two delegates who would attend the Edinburgh convention. Those chosen were Maurice Margarot, Chairman of the LCS, and Joseph Gerrald, an able barrister and prominent member of the Society who had just published a pamphlet urging that the House of Commons be replaced by a French-style 'Convention'.[26]

The Edinburgh 'convention' attended by Margarot and Gerrald was held a couple of weeks later. Its organiser was William Skirving, the son of a prosperous farmer from Midlothian, who had a wife and eight children. In publicising the event, Skirving had been careful to emphasise its delegates' peaceable and patriotic motives. Despite this, the convention ended abruptly when Margarot, Gerrald and Skirving were arrested in the middle of the night. The three were charged with sedition, found guilty and, like Muir, sentenced to transportation to New South Wales for fourteen years.

In April 1794, just before the prisoners were to depart for Australia in the transport ship *Surprize*,[27] the London Corresponding Society held another mass meeting, this time on the Green at Chalk Farm. An even larger crowd attended, James Parkinson among them, as one spy reported. After Thelwall and others had addressed the throng there was much discussion regarding the fate of the five 'Scottish Martyrs', as the prisoners were called, but other than raise funds for their families, there was little anyone could do to prevent their deportation.[28] Nevertheless, the size of this meeting and the alarming reports from the many spies present threw the Government into a state of panic. A few days later a bill was brought before the Commons restricting the holding of public meetings. William Pitt, in a long speech, spoke of a dangerous conspiracy being afoot – the country, he considered, was on the verge of catastrophe. Accordingly, he asked leave of the Commons to suspend *Habeas Corpus*, which meant suspects could be imprisoned and detained indefinitely.[29]

A couple of weeks later it was rumoured that the English reform societies were planning another convention and that an armed uprising was being considered. The Government, still in a state of high alert, reacted immediately. Early on the morning of 12 May 1794, Eaton's house was searched and plundered; had Eaton been present, he would have been arrested. He later recalled what happened: 'Ten officers and three of the city marshal men entered my house to seize me . . . on the Monday morning, and not finding me, on Tuesday they took up Mr Hardy' (Thomas Hardy, founder of the LCS).[30] Eaton does not say where he was but when the Bow Street Runners failed to find him, they instead summonsed his fourteen-year-old son, Harry, to appear before the Privy Council. Over the next ten days, thirteen prominent members of the LCS and six members of the SCI were arrested and imprisoned in the Tower of London, Thomas Hardy, John Thelwall and John Horne Tooke among them. This time the charge was treason, which meant that if found guilty the accused could receive the death sentence.

A few days later, on 25 May 1794, Parliament suspended the *Habeas Corpus* Act.

Without hesitation, James Parkinson stepped into the breach created by these arrests and accepted nomination to an Emergency Committee of the LCS, which would hold the fort until the prisoners were released.

5
The Pop Gun Plot

Treason: The highest crime of a civil nature of which a man can be guilty. In England, to imagine the death of the King is High Treason.

Webster's Dictionary, 1828

IT WAS AT this critical time that Parkinson first experienced an attack of gout. Having gone to bed free from pain, he woke in the middle of the night in excruciating agony, the pain resembling 'the stretching and tearing of ligaments or the gnawing of a dog'. The affected parts could not bear the weight of bedclothes, and even the tread of someone crossing the room caused him distress. He was mortified to discover that, like his father, 'I was also under the influence of this tormenting malady' at a relatively early age.[1] He believed he had a temperament that was prone to the disease because he was someone who was constantly 'distressed with anxious thoughts' and 'considerably engaged in study', although he recognised the main cause was that he had inherited it from his father, who had also had his first attack before he was 40.

Over the following months, James experienced increasingly severe attacks in his feet, hands and wrists, the pain often so relentless as to render 'the least attention to business distressing and irksome, to the highest degree'. As the attacks became more frequent he could sense one coming on from a general feeling of unease which was accompanied by numbness and coldness, and a sense of prickling in the feet and legs. During an attack he experienced 'the most distressing dejection of spirits, [my] mind being possessed with groundless apprehensions and alarms'. Both hands were badly affected, particularly the thumb and forefinger of his right hand, which must have made writing all those pamphlets and letters for the London Corresponding Society extremely difficult. Realising this makes his devotion to the cause all the more impressive.

Deprived of its Chairman, Secretary and eleven other leading members who had been imprisoned, the LCS was struggling to survive. Over the following months Parkinson attended meetings of the Emergency Committee almost every other evening, travelling the several miles into town for each one after a long day's work. They always sat until midnight, and often until two in the morning, despite some members having to get up for work at five o'clock the next day. Parkinson would then have to get home. He wrote and rewrote scripts for speeches and pamphlets, as well as attempting to procure articles from famous authors, such as the playwright Thomas Holcroft who supported the cause. He was even tasked with going to see the attorneys John Gurney and Felix Vaughan with a view to encouraging them to defend the prisoners and provide their services free of charge. In the event, two of the country's most prominent barristers, Thomas Erskine and Vicary Gibbs, agreed to counsel them, neither taking a fee for his services as it was considered unprofessional to take fees for defending people charged with High Treason.[2]

With suspension of the *Habeas Corpus* Act there was no pressure on the authorities to bring any of the cases to trial, so there was much concern that the prisoners would be incarcerated indefinitely and that their businesses and livelihoods would be ruined as a consequence. The Emergency Committee therefore established a fund to support the wives and children of the men in prison, and Parkinson wrote a penny pamphlet entitled *Revolutions Without Bloodshed*, the proceeds of which went to the fund.[3] In it he listed 24 benefits he considered a reform of Parliament would bring, many of which still resonate today:

- Taxes might be proportioned to the abilities of those on whom they are levied, and not made to fall heavier on the *poor* than the *rich* . . .
- . . . due PROVISION be made for the *aged* and *disabled* . . .
- Families that are comparatively starving might be exempted from contributing [taxes] towards the enormous sums squandered in *unmerited* SALARIES and PENSIONS . . .
- The EXPENSES OF THE NATION might not then *exceed*, as they do now, the enormous sum of £80,000 *a day* – £3,000 *an hour* – or £50 *a minute*.

But he had no real expectation that such reforms would be implemented by a Government 'disgraced by a *band* of men sent there by an established system of private patronage'. He ended the leaflet with a cry of 'TRAITORS! TRAITORS! TRAITORS!' which parodied a handbill being hawked about the town that contained accounts of conspiracies and treasons supposedly carried out by the LCS, and which was headlined TREASON! TREASON! TREASON![4]

Parkinson's highly provocative pamphlet was sold by Daniel Eaton and another bookseller, John Smith. By July, Smith reported his sales of the work totalled £3 16s, suggesting he had sold almost 1,000 copies.[5] Eaton probably sold a similar number, which meant that within a month of publication thousands of people across the city were familiar with its rousing message, having either bought it themselves, borrowed it from friends, or had it read to them in coffee shops and taverns.[6] Four months later, with the prisoners still incarcerated and no date set for their trial, there was a dramatic new development. An informer had reported to the authorities that several members of the London Corresponding Society, men who were 'still at liberty', were plotting to assassinate the King. This confirmed the Government's worst fears and plans were immediately set in motion for further arrests.

On the evening of Saturday 27 September, 1794, there was a loud knock at the door of a house where an eighteen-year-old watchmaker, Paul Lemaitre, was known to work. The door was opened by an apprentice who was immediately seized by the Bow Street Runners waiting outside. He was locked in a downstairs room with a Runner posted outside, pistol in hand. The rest of the troop, led by the 'Principal Officer' John Townsend, rushed upstairs where they found Lemaitre. Still in his work clothes, Lemaitre requested that he be allowed to change before being taken away, but 'I was only answered by oaths, and dragged down [the stairs] so forcibly that I was compelled to inform them I had some knowledge of walking alone'.[7] At the bottom of the stairs Lemaitre was surprised to see Thomas Upton lurking in the shadows, a man with whom he had had a significant disagreement a few days earlier, but when he called out to him to ask what was happening, Upton slunk away without replying. Several times Lemaitre demanded to know why he was being arrested and asked to see the warrant. Eventually Townsend put a pistol to his head and threatened to blow out his brains if he said another word. Lemaitre's hands were then tied and he was thoroughly searched and then escorted to prison while the remaining Bow Street Runners searched the house. As well as taking away a large quantity of papers, they seized a bag of marbles, a metal tube, and what appeared to be a gun. The tube was in fact the casing for a telescope Lemaitre was making and the gun was a toy made of tin; nevertheless, these items were removed as material evidence to support the case that an attempt was to be made on the King's life.

Later that evening, George Higgins was sitting behind the counter of the druggist shop in which he worked, when Townsend and three Bow Street Runners rushed in and arrested him, tying his hands with a handkerchief. Three more handkerchiefs were produced, into which they bundled a quantity of papers. The shop owner, on enquiring why Higgins was being arrested, was told he 'was engaged in a plot to furnish poison to kill the King'. Higgins was taken to Bridewell Prison in Tothill Fields and committed there two weeks later.

Next, it was the turn of the bookseller John Smith, but he was away on a fishing trip. He returned the following evening to find that his house had been ransacked and that papers and a large number of pamphlets – to the value of £14 and including Parkinson's *Revolutions Without Bloodshed* – had been taken away. Bow Street Runners were waiting for him and, without allowing him to change out of his fishing clothes, escorted him to Clerkenwell New Prison. There Smith was locked in a room 'which could hardly be born from the stench arising from a tub filled with excrements and urine'. The height of the windows prevented him from emptying the tub; there was no chair to sit down on and the bed was so damp that he did not dare take off his clothes. Two weeks later he was committed to Newgate Prison, 'on the felon's side'.* There the walls were 'naked, damp, and mouldy with one chair only and a broken table';

* There were two 'sides' to Newgate Prison, the felons' side, usually reserved for common criminals, and the state side. The relative luxuries of state side were only available to those who could afford to pay: a fee for admission as well as a weekly rent paid on a per bed basis. There was also an additional fee that could be paid for a private apartment. (Personal communication, Michael T. Davis, 11 January 2017.)

furthermore, there was no bed in the room until Smith was able to procure one from his own house.[8]

The first Parkinson knew about the arrest of his three friends was when he read about it in the newspapers on the following Monday morning. A fourth man, Robert Crossfield, was still being sought by the authorities. Apparently the informer was Thomas Upton, the man Lemaitre had seen hiding in the background during his arrest, but since then Upton had also been arrested because, according to the papers, he had 'prevaricated in his evidence, and been contradicted in material parts of it'.[9]

It seems Upton had informed the authorities that he and the other arrested men had conspired in a plot to kill the King but, unable to go through with this 'terrible deed', Upton had decided to turn King's evidence. He had told the authorities that the King was to be disposed of while he was in the theatre, by someone firing a poisoned dart at him from an air gun – which is why the case became popularly known as the Pop Gun Plot.

Parkinson's reaction to the news was coloured by the fact that he knew Higgins and Smith to be Upton's sworn enemies, so initially he laughed at the idea of these men collaborating on such a plot, given the relationship between them. Furthermore, the idea of killing the King in such a public place as a playhouse 'by an arrow, which was to be armed with a miraculous poison, and to be discharged from an air-gun, leisurely levelled for aim, in the midst of a crowded audience' was so ludicrous, 'I concluded such a story could not obtain a moment's credit, and expected your immediate discharge', he wrote in a letter to Smith. But when the discharge was not forthcoming and it became clear that the Privy Council intended to press ahead in its examination of the defendants in order to ascertain whether or not there was a case to answer, Parkinson quickly realised just how serious the situation was. His letter to Smith continued: 'I endeavoured in vain to chase from my mind the idea that you and your fellow-prisoners were to be held out to the world as liable to be HANGED, DRAWN, and QUARTERED, for your *imputed* crimes.' All detainees were charged with High Treason, an extremely serious offence; if convicted each was liable to be hanged by the neck, cut down while still alive, disembowelled, his entrails burned before his face, and ultimately beheaded and quartered.

Gradually it dawned on Parkinson that the authorities might seize this opportunity to make examples of his friends. If they could prove there really had been a plot to kill the King, the success of such a trial would undoubtedly benefit the Government by turning public opinion against the radical societies. For although there was widespread support for the French Revolution, the average Englishman stopped short at the idea of regicide and had been genuinely shocked by the barbaric deaths of Louis XVI and Marie Antoinette at the hands of the mob. Consequently, if the public could be convinced that the reform societies really were plotting to kill the King – an accusation they had constantly denied – it would lead to members leaving in droves. Furthermore, it would have serious repercussions for the men imprisoned during the first wave of arrests who were still waiting to come to trial, as their juries might be led to believe 'that TREASON was actually abroad'.[10] An added difficulty, Parkinson realised, was that the charge of High Treason not only included the *act* of killing the King, but also just *imagining* his death, so if the authorities were able to convince a jury that this had been a serious plot, even though it had not actually been carried out, there would be no hope of reprieve for the prisoners.

Convinced that Smith, a man he had known for a number of years, was innocent, Parkinson at once determined that the Privy Council should be informed of the ludicrous nature of the accusations and immediately had them informed that he was willing, under oath, to put his version of events before them. Parkinson had witnessed the events over the preceding few weeks that had caused Upton to fall out with Smith, Higgins and Lemaitre, so he believed that if could explain these to the Privy Council they would understand why Upton had made up the accusations. In the prevailing political climate, stepping forward in this way was an extremely brave and indeed risky thing to do. The Privy Council, which included the Prime Minister and his Secretaries of State, constituted the highest court in the land and Parkinson must have realised that if he could not convince them with his testimony, he ran the risk of being implicated in the plot himself. He would be putting his own life in danger. Indeed, as soon as the newspapers reported that Parkinson was to be examined by the Privy Council, rumours abounded that he too was involved in the 'horrible plot'.

Parkinson had met Thomas Upton for the first time only a few weeks previously, and then in rather unpleasant circumstances. A member of the LCS Emergency Committee had heard that Upton had been indicted for setting fire to his house in order to defraud his insurance company and that he had only escaped prosecution because of some loophole in the law. Since Upton was one of nine LCS members entrusted with collecting funds for the imprisoned members' families, this rumour of his dishonesty could not be ignored. Parkinson, shocked by the revelation, had insisted that the matter be investigated, so John Smith and George Higgins were asked by the committee to make further enquiries. In the meantime, Upton was sent a letter from the committee requesting that he remove his name from the list of people collecting money while enquiries about his activities were ongoing.

Unknown to Parkinson and others on the committee, it seems that Upton had been tasked by a magistrate to infiltrate the London Corresponding Society and initiate a plot to kill the King, as that would give the Government reason to imprison senior LCS figures who had stepped in when the leaders had been imprisoned, and would also drive away their supporters. Government spies abounded in the reform societies, planted by magistrates who had been instructed by the Home Office to gather information on the societies' activities. Although some spies were patriotic volunteers, many were common criminals who offered their services in return for getting out of jail, so their evidence could be highly suspect, sometimes simply being invented in order to convince the magistrates they were keeping their part of the bargain.[11] As Upton was evidently a somewhat nefarious character, it is possible that some deal was done whereby the magistrate provided Upton with the loophole regarding the prosecution over his attempt to defraud his insurance company, in return for Upton's role in initiating a plot within the LCS to kill the King. What is known for certain is that Upton was in the Government's pay at this time.

The surgeon Robert Crossfield – a dissipated character who took large quantities of opium – had been a member of the LCS for some time. Upton had befriended him at an LCS meeting, perhaps recognising in Crossfield the kind of dissolute individual he needed to help implement his plans. In his role as *agent provocateur*, Upton embroiled Crossfield in his scheme to kill the King, which Crossfield seems to have gone along with very willingly. It would be several years before Crossfield's role in the Pop Gun Plot was revealed, as he went on the run when Smith and the others were arrested, escaping to sea, but when he was finally captured and put on trial, a number of people reported how they had seen Crossfield and Upton together making enquiries about the construction of a brass tube. A diagram was produced at the trial, apparently drawn by Crossfield, which showed how this tube would be used to construct the air gun from which the poison arrow would be shot. Other witnesses, particularly sailors on the ship Crossfield had absconded in, related how he had bragged that he had been the ringleader in the plot, had designed the airgun and prescribed the poison to be used on the arrow. They also told how he claimed to have already tried to shoot the King in his carriage in Buckingham Palace Road, 'but unluckily missed', and how he would still like to assassinate the King and the whole damned Government. Sometimes calling himself Tom Paine, it seems probable that in his opium haze, Crossfield found it hard to tell fact from fiction.[12]

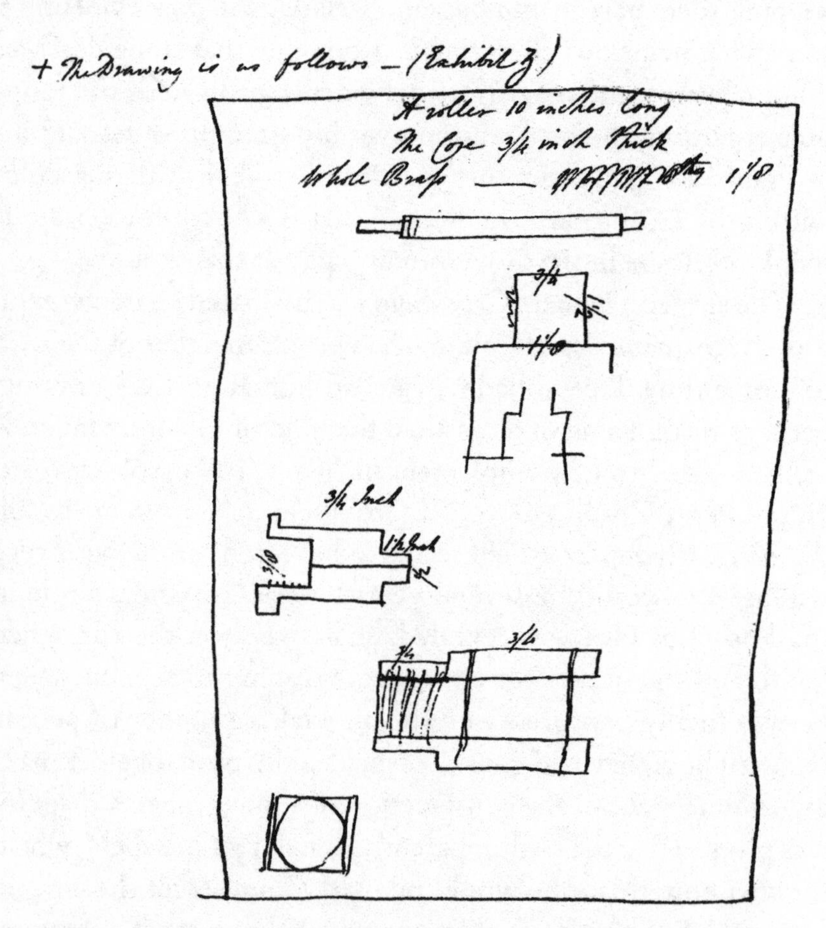

A sketch of the brass tube from which the 'pop gun' was to be made. This was used as Exhibit Z in Crossfield's trial.

When John Smith and George Higgins finished their investigations into Upton's fire, both reported that Upton's neighbours were convinced he had started the fire deliberately; furthermore, Smith and Higgins heard many other 'disagreeable imputations' against Upton. This led them to recommend that Upton be removed not only from the list of subscription collectors, but from the Society altogether. When Upton got to hear about this, he was outraged. He went to see Smith in a furious temper and demanded the names of the committee members who had recommended his expulsion. Smith refused to give this information, and advised Upton to provide a written reply at the next committee meeting, but instead of doing this Upton sent an anonymous letter to the committee under the pseudonym 'Pasquin'.

At the next meeting, on 4 September, the anonymous letter was read aloud to members. It lectured on the evils of bringing charges against members without adequate evidence, and charged the committee with 'commending such evils'; thus it was immediately evident that the author was Upton, who was in the room. Angered by this attempt to anonymously discredit the committee in front of its members, the outspoken young Lemaitre declared that Upton's behaviour showed he was unworthy of being called 'citizen' and that he should be removed from the Society immediately. Hearing this, Upton became extremely angry 'giving him the appearance of a madman', and breaking out into 'a strain of abuse'. Following Higgins's suggestion that Upton should be censured for this behaviour, Upton's anger became so ungovernable he declared he would 'never rest satisfied' until he had revenged himself and obtained satisfaction for the insult he had suffered. But sensing he had overstepped the mark and that the committee was unanimously against him, and wishing to avoid the disgrace of being censured, he was making for the door when Higgins rose to inform everyone that if they were going to censure him they had better be quick about it as he appeared to be 'hopping off'. Upton, who was lame with a club foot, took this as a personal insult. He turned at the door and yelled that Higgins would repent his words 'before many weeks are over'. It was no idle threat.

The debate at the next committee meeting regarding Upton's extreme behaviour had again led to a motion for his expulsion from the Society. However, after further discussion, during which Parkinson argued that perhaps their first letter to him had been too harsh, since they did not have actual proof of his guilt, Parkinson agreed to write to Upton personally. In this letter, which he hand delivered to Upton's house, Parkinson repeated the suggestion that Upton withdraw his name from the list of collectors, but explained that this was not because the committee had charged Upton with any crime, but because they were concerned that his name on the list would lessen the number of subscriptions. Upton wrote back two angry letters. In one he again asserted his innocence; in the other he challenged Lemaitre to a duel with pistols. Lemaitre immediately replied, rejecting Upton's challenge as childish, but in the light of what happened next, he later suspected that Upton's letter had been sent merely to prompt a reply, so Upton could obtain a specimen of Lemaitre's signature and handwriting.

At Crossfield's trial, it would emerge that Upton had forged two letters, ostensibly written by Lemaitre; one was addressed to Smith and the other to Higgins. These letters described the tube from which an air gun was to be built, giving its dimensions and weight. The letter to Smith also contained a plan of how it would be built. This damning evidence made it appear that it was Smith, Higgins and Lemaitre who had been plotting to kill the King, when in fact they had had nothing at all to do with it. They had simply become caught up in a personal vendetta when Upton realised he had an opportunity to revenge himself on those who had insulted him.

When Upton received Parkinson's letter he realised that despite its conciliatory tone, he was likely to be expelled from the Society at any moment, so decided now was the time to make his move; if he delayed until he was expelled he would be unable to complete the task with which he had been charged by the magistrates, and would therefore not get paid. So the next day he took the forged letters to the authorities and reported that Lemaitre, Smith, Higgins and Crossfield had ordered the construction of an air gun, and that they planned to use it to shoot poison arrows at the King with the intention of assassinating him. He further asserted that the person who had initiated the idea was Higgins, and that it was he who had asked Upton and Lemaitre to make a metal tube that could be used as an air gun. This was to be disguised as a walking stick and loaded with a poison dart, the poison being a rare and deadly substance which Crossfield or Higgins would supply. He argued that as Crossfield held a medical degree from the University of Leiden and Higgins was an apprentice chemist, between them they had considerable knowledge of poisons. The plan, according to Upton, was to smuggle the gun into a theatre when the King was in attendance: at some point a disturbance would be created among the crowd at floor level and when the King leaned forward from the Royal Box to see what was happening, he would be shot with the poison dart. The culprit would then escape under cover of the ensuing commotion.

The next day Lemaitre, Smith and Higgins were arrested and, as the only person really involved in Upton's plot, Robert Crossfield went on the run.

6
Trials and other tribulations

'To whom can we look for redress? To Ourselves. Generally suffering, let us unite for general relief.'

Old Hubert, 1795
Ask and You Shall Have

SHORTLY AFTER the arrests of the Pop Gun suspects, Parkinson was at a meeting of the Emergency Committee in Chancery Lane when the door suddenly burst open and to everyone's amazement Thomas Upton rushed in.[1] According to the newspapers, he was supposed to be behind bars. Upton was closely followed by several Bow Street Runners, who were looking for Richard Hodgson, an active member of the committee, but as Hodgson was not present they arrested Burks, the Secretary, instead and seized all the papers that lay on the table. As they were about to leave, Upton turned and pointed an accusing finger at Parkinson, exclaiming: 'If there are any papers of consequence, they are about that man.' Parkinson was duly searched and a number of papers found in his pockets were taken away.

Raids like this were becoming a regular occurrence, so after the Runners left with Burks, the committee hastily adjourned to a public house in Butcher Row in case they should come back for anyone else. While there they collected six shillings for the relief of Mrs Burks who was not likely to see her husband again for some time to come.[2] The following day Parkinson received a letter from the Privy Council summonsing him to attend the Council chamber in Whitehall at '10 o'clock on Tuesday 7th of October', in order that 'you may be examined, and give your testimony on certain matters now pending before the board'.

Parkinson duly arrived at the appointed time, but was required to wait half an hour in the antechamber while their Lordships were engaged on other business. Upton was also there and Parkinson presumed he too was waiting to be questioned, but after a few minutes someone came out of the Council chamber where the case was being heard and asked for a messenger to fetch Upton's wife, who was required as a witness. When no messenger could be found, Upton himself was told to get her. Parkinson could not believe what he had just witnessed: 'I saw this man, the accuser of three men of High Treason, and himself a supposed conspirator against the life of the King, leave the room to walk the streets *alone*, and confer with whomever he might choose.' This only served to confirm Parkinson's suspicions that Upton was a government agent.

Eventually Parkinson was called in to the Council chamber. Sitting around a large table were the Prime Minister William Pitt the Younger, the Home Secretary the 3rd Duke of Portland, and a number of other lords, earls, dukes and bishops who made up His Majesty's Lordships of the Privy Council. Presiding over them all, sitting on a raised, high-backed chair, was the Attorney General Sir John Scott.[3] Various ladies were seated in the audience and other interested parties looked down from the balcony above. It was an imposing and forbidding occasion but Parkinson stood confidently before them, making no attempt to ingratiate himself. First he refused to take the oath unless he was informed 'on what points I am to be examined'. When Pitt told him that that was impossible, Parkinson explained that while he was happy to be examined respecting the business of Smith, Lemaitre, Higgins and Upton, he would decline to take the oath regarding any other matters. Irritated by Parkinson's lack of cooperation, the Attorney General reprimanded him: 'Consider, Sir, you are now before the *highest court* in this kingdom.' But Parkinson persisted until Pitt pointed out that Parkinson had been called to answer questions respecting 'matters of the highest importance to the state', and while the intention was to examine him on the business of Mr Upton, it was nevertheless possible that questions might arise that did not immediately appear to be relevant. On these grounds Parkinson graciously conceded to take the oath on condition that he would be allowed to object to questions likely to 'criminate' him.

The examination commenced with the Prime Minister asking Parkinson how long he had known John Smith. 'I believe about two years,'[4] he replied, but as he started to elucidate this statement, Pitt interrupted, asking him to speak slower so the clerk could take down his words.[5] Parkinson responded by advocating an alternative approach: 'If you please, Sir. But I think I could suggest a better mode of examination to their Lordships.' Pitt was astounded at his audacity: 'What – what is that?' he shouted. Parkinson replied, 'That your Lordships will allow me to give you an uninterrupted detail of what I know . . . I am sure it will save both your Lordships and myself much time and trouble.' Parkinson's fearlessness when confronted by the most formidable individuals in the country is astonishing. To his credit, Pitt saw the logic in this argument and acquiesced: 'Very proper, Sir; we shall be much obliged to you.'

Parkinson went on to tell them what he knew about the alleged plot, explaining the grounds of Upton's grievances which he believed had caused Upton to make up the accusations against

Smith, Higgins and Lemaitre, and pouring scorn on the ludicrous nature of such an unlikely conspiracy:

> A poisoned arrow was to destroy the King, and to what purpose? would not another King succeed; or if instant death had not followed the unerring dart of wonder-working poison, to what advantage to any conspirators could have been the illness of a King? Has he not been for several months unable to hold the reins of government yet the nation was not, in consequence of his madness, in a state of anarchy and confusion? What could the plotters gain by this attempt?[6]

An engraving of Benjamin Franklin before the Privy Council in 1774 gives a good impression of what it must have been like for Parkinson to be interrogated by them.

When he finished his explanation, the interrogation began. Immediately he was picked up on a point he had made about wishing to rid the LCS of members with bad moral characters. Who had he meant? Parkinson made it known in no uncertain terms that he had been referring to members who were government spies. In particular, he had in mind two Emergency Committee members, John Groves and William Metcalfe, both of whom Parkinson had worked with on many occasions. Metcalfe often chaired meetings and had even presided over the one at which Groves had been accused of being a spy. But it was difficult to prove such things and in the end Groves was acquitted of the charge, 22 voting in his favour with only four against – but Parkinson had clearly not been convinced. He also suspected that Metcalfe was a spy, but it nevertheless occurred to him that here was someone who could corroborate his story and confirm his friends' innocence. Metcalfe had been present at all the relevant meetings and would therefore know that there had been no discussions about assassinating the King; furthermore, he was more likely to be believed, since he was in the Government's pay. So Parkinson casually remarked that he was surprised the Privy Council had not already questioned Metcalfe since, he observed with disdain, Metcalfe already had 'very easy access to your Lordships'.*

As he spoke those words, he handed the Attorney General Metcalfe's business card, so he could have his address. 'Is this your hand-writing, Mr Parkinson, at the back?' enquired the Attorney General as he examined the card. Parkinson froze; he had forgotten writing on the back but immediately understood the significance of the question. He knew that the authorities were in possession of many papers in his handwriting and now the handwriting on the

* Metcalfe was indeed another spy, as revealed in a letter from him to a magistrate, 5 January 1795. (National Archives, Kew, HO 42/34/6.)

back of the card could identify him as the author of this material. He nevertheless put on a brave face and with a smile 'more expressive of contempt than satisfaction', replied, 'It is.'

Eventually the questioning turned to the whereabouts of Richard Hodgson, whom the authorities had been seeking when they burst in on the Emergency Committee meeting at which Parkinson had been searched, and who had still not been found. Hodgson should have been arrested along with Thomas Hardy and the others during the first wave of arrests but he had continued to elude the authorities and was now indicted on a charge of High Treason. Suddenly the Attorney General turned to Parkinson and asked when he had last seen Hodgson. 'Last night,' came Parkinson's prompt reply. A gasp of surprise went up. '*Where* did you see him?' barked the Attorney General. Parkinson objected to the question.

PITT: What do you mean, Sir? You must answer the question.

PARKINSON: This question, more than many others I have already answered, can have no reference to the pretended plot.

PITT: That does not signify; you are bound by the oath you have taken to answer it.

PARKINSON: But it was surely agreed that I was to object to certain questions.

ATTORNEY GENERAL: Yes, to such as might criminate yourself, and to those only. Choose whether you will answer the question or take the consequences of refusal. Now, Sir, where did you see Hodgson?

PARKINSON: I wish not to behave with incivility towards your Lordships, but I must say . . . that I am used exceeding ill.

PITT: Sir, you cannot object to this question.

PARKINSON: I conceive that I *can*, and *do* on this ground also. That you ought not to put such questions, the refusing to answer which will imply crimination.

ATTORNEY GENERAL: Sir, you must answer the question.

PARKINSON: My Lords, my legal knowledge is but very trifling, it chiefly consists in knowing what was a crime a few years ago [i.e. before the Government's clampdown on political activities]: but from the extraordinary circumstances I have lately observed, I know not what may now be decided a crime or not. On that ground also I object to answering this question.

ATTORNEY GENERAL: Then, Sir, you know that a bill was found yesterday for High Treason against Hodgson, by a Grand Jury of his countrymen?

PARKINSON: I learned so from the public papers, and indeed, from himself.

ATTORNEY GENERAL: Now, Sir, if after that you conceal him, you are guilty of Misprision of Treason.*

PARKINSON: That, Sir, I am aware of.

ATTORNEY GENERAL: Now, Sir, answer the question; but you are not wished to criminate yourself.

PARKINSON: Then, I saw him in *my own house*.

A ripple of surprise went through their Lordships.

ATTORNEY GENERAL: At what time?

PARKINSON: About eleven in the evening.

ATTORNEY GENERAL: What did he come about?

PARKINSON: Nothing particular; he laughingly told me of the bill found against him.

ATTORNEY GENERAL: Do you have visitors ever at that late hour?

PARKINSON: Yes, often my Lords, I sit up late.

* Misprision of Treason meant that Parkinson knew Hodgson intended to commit treason, but did not tell the authorities.

Gradually the Attorney General managed to drag answers from Parkinson who was eventually pressed to say that before calling on him, Hodgson had been in a public house in Shoreditch, one facing the church that he thought was called either the *King's Arms* or the *Star and Garter.* Having divulged this information, Pitt scribbled something down on a piece of paper, probably the names of the pubs, and took it out of the room, presumably to instruct someone to search them for Hodgson. The questioning then turned to some pamphlets the Attorney General was holding.

'Pray Sir,' said the Attorney General, 'did you ever see one of these pamphlets?' He held up a copy of *A Vindication of the London Corresponding Society* which had been written by Parkinson for the LCS a few weeks previously, although the work did not identify him as the author. On 6 September 1794 *The Times* had devoted a whole page to what it called *The New Times.* This looked exactly like a normal page of the paper, but every article was in fact a satire of life in England as it might be after a revolution. In *A Vindication* Parkinson complained that the satire was a violation of the law, for it assumed Thomas Hardy and the others accused of treason were guilty before they had been tried; it was a fundamental principle of the constitution that they be declared innocent until proved guilty.

Fear for the lives of the prisoners had considerably increased following a recent treason trial in Scotland in which the two defendants had been found guilty – one of them being sentenced to death and the other to exile.[7] *A Vindication* also castigated the prosecutor for the Crown in that trial, John Anstruther (shortly to be Sir John), and essentially accused him of lying. *The New Times* had been published just weeks before the English treason trials and Parkinson was concerned that not only was it an inaccurate record of the London Corresponding Society's activities, but that it might also influence the jury against the defendants.

He accused *The Times* of publishing in the full knowledge that it would not be prosecuted, since it supported the Government's position.

It was therefore with some trepidation that Parkinson replied to the Attorney General's question, saying that he was indeed familiar with the pamphlet.

'Do you know the author?' asked the Attorney General.

'Exceedingly well,' replied Parkinson.

'Who is the author?' inquired the Attorney General again, but this time Parkinson objected to the question. One of their Lordships turned to the clerk saying: 'Put down "Refused to answer because it will criminate himself".' Parkinson struck his fist on the table in anger and declared, 'My Lords, I claim that no words be inserted . . . but what proceeds from my own lips – I gave no such reason.' Another of the Lordships redirected the clerk: 'Only put down "Refuses to answer question".' More compliantly, Parkinson proposed that if the question were put again he would answer it. The Attorney General asked once more: 'Who is the author?' Parkinson replied quietly: 'I am the author.' '*You* are the author?' roared the Attorney General. 'I am the author,' repeated Parkinson defiantly.

ATTORNEY GENERAL: And pray, was this intended to be published before the [treason] trials?

PARKINSON: It was – But on its being suggested, I believe by myself, that it might not be proper, it was resolved it should not be, the press was therefore stopped.

ATTORNEY GENERAL: And how many were printed then?

PARKINSON: Only two hundred and fifty.

ATTORNEY GENERAL: How many had been intended?

PARKINSON: I think two thousand.

PITT: Pray, Mr Parkinson, how came you to be invited on this Committee [the LCS Emergency Committee]?

PARKINSON: Because, I believe they did me the honour to believe me firm in the cause of parliamentary reform. [Here, said Parkinson later, Pitt smiled, presumably remembering when he too had been a firm supporter of parliamentary reform.] And because I had just published a little tract for the benefit of the wives and children of those imprisoned for High Treason.

ATTORNEY GENERAL: Pray what was that?

PARKINSON: *Revolutions Without Bloodshed*.

ATTORNEY GENERAL: That was yours too?

PARKINSON: It was.

Quite why Parkinson volunteered this information is unclear, as the first part of his answer would have been sufficient. Perhaps he got carried away by a sense of bravado, but of all his publications *Revolutions Without Bloodshed* was the most seditious with its heartfelt cry of 'TRAITORS! TRAITORS! TRAITORS!', so he was treading on extremely dangerous ground. Nevertheless, the questioning eventually came to an end and he was allowed to go home – although rumours abounded that he had been clapped in irons and sent to Newgate Prison; a relation even came to the house to offer condolences to Parkinson's wife.

Since Parkinson had certainly 'criminated' himself regarding his political views and activism within the London Corresponding Society, and also admitted to writing two highly inflammatory and seditious pamphlets, it is rather curious as to why he was not incarcerated there and then; many had been imprisoned for far less. Perhaps the Privy Council realised he was not directly involved in any plot to assassinate the King, so any move to imprison him after he had volunteered to put his evidence before them would have risked making a martyr of him. Parkinson was a popular and well-known figure and his arrest might have led to a surge of sympathy for the reform movement; perhaps a more circumspect decision was not to apprehend him.

Another reason might be that he was only the author and not the publisher of such works, for the latter were considered far more dangerous when it came to 'spreading poison', as evidenced by the prosecutor at one of Daniel Eaton's trials who accused booksellers such as Eaton of dealing in a 'very dangerous commodity':

> Mr Paine shall have my consent to sit down and write till his eyes drop out and his heart aches, provided he cannot find any body to publish it; but it is by means of persons like the defendant, giving vent to publications like the present, that injury has been done to society.[8]

Whatever the reason, Parkinson's lucky escape appears to have emboldened his spirit, for a couple of months later an advert in *The Times*, listing several of his publications, included 'Revolutions Without Bloodshed, and the other Political Trifles of Old Hubert. By JAMES PARKINSON.'[9] Having been forced to disclose his alias while under interrogation, he was now using that exposure to his advantage.

Shortly after his encounter with the Privy Council, Parkinson learnt that all his efforts to prevent Smith and the others going to trial had been to no avail, for the Privy Council had decided that there was sufficient evidence to proceed with their prosecution. For months the prisoners were detained in appalling conditions and given no indication of when their cases might be heard. The winter of 1794–95 was exceptionally harsh, with the bitterly cold weather beginning on Christmas Eve. January 1795 was the coldest month since instrument records had begun in 1659, -21°C being recorded in London on 25 January. The Thames (and the Severn) froze over and 'Frost Fairs' on the ice started up again after an absence of several years when the weather had not been cold enough. In early February there was a rapid thaw and flooding ensued. The severe cold then returned mid-February and continued well into March.[10]

With no heating in Smith's cell, only paper covering the windows, and few blankets to wrap himself in at night, it is not surprising that in February 1795 Smith sent Parkinson a letter, begging him to come and see him as soon as possible. 'A damp room and cruel treatment is so injured my constitution that I've been confined to my bed for some time dangerously ill.' Parkinson went immediately and was appalled by what he found. He wrote instantly to the Privy Council, complaining of the disgraceful conditions in which Smith was held, stating that unless he received the care and attention he needed, he was likely to die.[11] Four days later, Smith was moved to a more tolerable cell and a physician appointed to attend him.

Lemaitre fared little better. He subsequently wrote an account of his arrest and treatment in Newgate Prison – which he called the British Bastille – in which he related the dreadful conditions he had been subjected to and the death of his mother whose illness, he claimed, had been hastened by the distress of her son's imprisonment. He had suffered a constant pain in his side, which baffled the 'Bastille Doctor' but which Lemaitre believed was caused by his bed being so near the damp walls. Furthermore, his 'cave' had no bedstead (presumably his mattress lay on the ground), no boards on the floor, nor plaster on the walls. He had repeatedly asked if he could be moved to better conditions, and it was only when his mother died that his request had been granted. He was then moved to a cell where he had a view of the countryside and could see in the distance 'my fellow citizens walking without the (visible) shackles of bondage'.[12]

At the end of April 1795, when the prisoners had been incarcerated for seven months and were still without a date for their trial, Parkinson was again summonsed to Whitehall. He was informed that as Crossfield still could not be found, Pitt had decided to release Higgins and Lemaitre on bail. Smith, however, had greatly irritated the Privy Council by writing to them in an aggressive manner, threatening to publish an account of his terrible treatment in prison if he was not immediately brought to trial or allowed out on bail. It was therefore suggested to Parkinson that he should speak to Smith and tell him that if he wrote again in more 'respectful terms', then he too would be released on bail. Although this is what eventually happened, Parkinson privately railed against having to ask Smith to demean himself in this way.

Accordingly, on 9 May 1795, Smith and Lemaitre were released on bail following the payment of 'one guinea and half-a-crown' and giving assurances regarding their future appearance at the trial. They were required to give surety of £200 for themselves and find two others prepared to pay £50 each if they did not appear. Higgins was released under the same conditions on 18 May. Three days earlier Samuel Newport, Keeper of the New Prison Clerkenwell, had received a note signed by William Pitt, the Duke of Portland, and six other members of the Privy Council, ordering Newport to discharge Thomas Upton from his confinement in prison because he too was being let out on bail.[13]

Parkinson continued to write political tracts throughout 1795 still using the name of Old Hubert, some of which were extremely vocal in their attacks on the Government, particularly with regard to the worsening food shortages. A poor harvest the previous year, followed by the severe winter, was then followed by another poor harvest. That, coupled with imports being blockaded by the French, with whom the country was still at war, led to a serious shortage of wheat and other commodities. Prices rose rapidly, workers were laid off, wages were cut and many people were on the brink of starvation. The result was a wave of food riots that spread across the entire country. Parkinson continued to complain vehemently against a corrupt and evil Government which he believed was the cause of these problems. The poignant opening lines of one of his pamphlets outlines the situation:

> The hour of Calamity is arrived; Famine is already at our doors. The schemes of an evil ministry have at last brought us to really know the want of food. In their endeavours to enslave Frenchman, they have wasted the Bread which should now feed us and our children . . .

'God save the people!' he cries forlornly at the end.[14]

This pamphlet was followed by two others on food shortages; his anger and frustration mounting at the gross injustices suffered by the poor. *Whilst the Honest Poor Are Wanting Bread* details a fictitious conversation between father and son, overheard by the narrator.[15] The two men, unable to find employment, have taken a stroll to 'hide their suffering from the tender part of their family', and 'so the women and young folks will have more to eat by our going without a dinner'. On their walk the young man picks up a tattered fragment of newspaper and reads it to his father 'to withdraw your thoughts from our present state of distress'. But the newspaper is full of examples of the profligacy of the upper classes: the mistress of a house who spent '£20,000 on champagne to wash with'; the gentleman who paid 'seven guineas a pair!!' for boots; the dog kennel that was 'convincing proof of his Grace's taste in architecture';[16] and the poor Princess of Wales who had to be supported on her way to the chapel owing to the 'weight of her ornaments and jewels being so excessive'. 'And all this,' says the old man, 'WHILST THE HONEST POOR ARE WANTING BREAD.' Although the story is fabricated and highly sentimental, as was the fashion of the time, it illuminated the vast divide between rich and poor that made Parkinson prepared to risk his life for the cause of parliamentary reform.

Robert Crossfield was eventually captured in December 1795 when the Government issued a £200 reward for his arrest.

After going on the run he had offered his services as a surgeon on board a ship bound for the South Seas to hunt whales, but during the course of this voyage the crew had been captured by a French corvette – a small warship – and the English prisoners taken to Brest in France. Having endeared himself to his French captors by convincing them that he supported the Revolution, Crossfield had been allowed ashore on occasions to meet members of the Convention. Unfortunately for Crossfield, even in the eighteenth century, international law provided protection for ships employed on humanitarian voyages, such as those carrying prisoners for exchange. When one of these cartel ships arrived at Brest to collect prisoners, Crossfield was inevitably returned to British shores.

On landing in England, a fellow prisoner who had witnessed Crossfield's fraternising with the French and heard him brag about his plans to kill the King (and perhaps putting two and two together when he learnt of the reward offered for Crossfield's capture), had reported his conduct to the authorities, whereupon Crossfield was apprehended. Even then Crossfield tried to escape by bribing his captors to let him go, but the proffered five shillings was clearly insufficient.

A month after Crossfield's arrest, Smith, Higgins and Lemaitre were again apprehended pending their trial, although it was not held until the following May, so they spent another five months in prison. Crossfield was the first of the four to be tried, on 11 May 1796. On the first day no fewer than seventeen people were called to give evidence against him, most of them witnesses to his incriminatory bragging. Parkinson was called as a witness for the defence on the second day and again explained why he thought the 'plot' had been fabricated by Upton who had a grievance against members of the London Corresponding Society. He was cross-examined by the Attorney General who clearly remembered him from their previous encounter, but again Parkinson had no fear of the man's authority and was quite prepared to stand up to him:

> ATTORNEY GENERAL: You had a considerable situation, I understand, in the society you belonged to – you were one of what is called the Committee of Correspondence?
> PARKINSON: I belonged to the Committee of Correspondence.
> ATTORNEY GENERAL: Sometimes, I believe, called the Secret Committee?
> PARKINSON: Once called so, in my hearing, by Upton, for which he was very much reprobated.

Upton, however, was not present. Instead, his wife was there, dressed in widow's weeds, claiming that he had committed suicide by drowning himself in the Thames, despite the counsel for the defence tendering two witnesses who said that Upton was alive and at his residence in London on the morning of the trial. The Court, however, inexplicably ruled this evidence as being immaterial to the case.[17] Despite the evidence against Crossfield, without Upton the trial only lasted two days and after deliberating for less than two hours, the jury had no hesitation in returning a verdict of 'Not Guilty'. Crossfield was immediately discharged. A week later, Smith, Higgins and Lemaitre appeared briefly in court only to be told by the judge, 'I am satisfied, as well as I can be . . . that that man [Upton] is not in existence.' With the prosecution unable to produce their main witness, the case collapsed and all four men were allowed to go free.

While Smith and the others were in prison, the trials of all those caught up in the first wave of arrests for High Treason, including Thomas Hardy, John Thelwall and John Horne Tooke, had gone ahead. But to a man all thirteen had been found 'Not Guilty', much to the embarrassment of the Government. During those trials, Groves and Metcalfe were exposed as government agents charged with spying on the London Corresponding Society; it therefore seems probable that Upton's death was fabricated by those in power, in case his true role in the whole sordid story came to light during Crossfield's trial and humiliated the Government even further.

Upton's 'confession' to the authorities had resulted in three innocent men being incarcerated for more than a year in the most appalling conditions for a crime they had not committed, and the hot-headed young Lemaitre was understandably furious about it. So when, several months after the trial collapsed, he saw Upton in Holborn – evidently still very much alive – Lemaitre endeavoured to arrest him with a view to delivering him to justice. Unfortunately, a riot ensued and Upton 'hopped off' in the confusion. Although Lemaitre reported this incident to the Attorney General and William Pitt, no action was ever taken.

Almost 200 years later, found among a bundle of accounts sent in 1798 from the Treasury Solicitor to the Home Office was a small but highly significant entry: the payment of £69 6s to Thomas Upton, 'For the maintenance of his family' between 26 September 1794 and 25 May 1795.[18] These dates exactly bracket the time Upton was in prison and the money was evidently payment from the Government to compensate his family for the eight months that he was unable to support them due to his confinement. The amount paid, however, was derisory when compared to the £300 a year paid to Metcalfe for his spying activities, which did not include a term in prison.

Altogether, five members of the reform movement had been unjustly transported to Australia for sentences of up to fourteen years; one member had been executed, another exiled, and dozens of authors, printers and booksellers had languished in prisons under reprehensible conditions, all to no avail. In the end, the Government never produced any evidence that there had ever been a serious plot to foster an uprising or to assassinate the King. Despite this, the authorities were still determined to suppress the reform societies and new laws would soon place even more severe restrictions on their activities.

True to his word, when John Smith was finally released from prison, he wrote an account of his and Higgins's arrest, imprisonment and trial in a pamphlet entitled *Assassination of the King!*. In this Smith detailed everything that had happened to him and the other accused during the twenty months from their arrest in September 1794 to their release after the trial in May 1796. He also asked Parkinson, 'that worthy citizen who did not scruple to come forward in the midst of danger' to recount his interrogation by the Privy Council and subsequent events, which Parkinson did in some detail in his 'Letter to Mr Smith'.[19] Parkinson ended with a call for a public enquiry and compensation for the defendants' 'loss of time, loss of property, and loss of health' during their time in prison. Needless to say, they got neither.

A few months later Smith was arrested again, this time for selling *A Summary of the Duties of Citizenship!* The pamphlet had been written in October 1795 by an LCS member, Edward Iliff, 'expressly for members of the London Corresponding Society'. It contained a message similar to many other LCS publications, but a particularly contentious paragraph appeared to specifically threaten the King:

> Unfortunately for us Englishmen, every war here must be a King's war and the people must support it! fight for it! die for it! – Is it not fair then to say, Kings should be the only sufferers? If blood be necessary, should not their blood be spilt, and in their own dispute too?[20]

Five days later, as the King was in a carriage on his way to open Parliament, he was apparently attacked. Even though the most likely explanation of this incident was that someone had slipped off the pavement and accidentally fallen against the carriage, the

LCS was widely blamed for the attack and for having stirred up dissatisfaction among the populace. Two proclamations were promptly issued, one offering a reward for finding the King's assailant and the second against seditious activities. The latter alluded to a recent LCS meeting at which 'divers [*sic*] inflammatory discourses were delivered to the persons so collected'. Between October and December 1795 the LCS had organised meetings that were the largest the city had ever seen. John Thelwell, Richard Hodgson and other treason trial 'celebrities' spoke to the crowds, which must have been a major attraction. The numbers that gathered in the fields of Islington and Marylebone and surrounded Parliament were estimated to be almost 300,000.[21] The Government, terrified by the size of these demonstrations supporting the LCS, responded savagely and introduced two new Acts: the Seditious Meetings Act and the Treason Act, which became known jointly as the 'Gagging Acts'. Meetings of more than 50 people could not be held without notifying a magistrate and speeches 'inciting people to hatred or contempt' of the King or Government became punishable by death. Defiance of bans on meetings was also punishable by death.

Smith again languished in prison for some months before being brought to trial in December 1796. This time, despite being defended by John Gurney, he was convicted of publishing 'one of the most villainous and atrocious libels that ever appeared in a Court of Justice'. And although both a 'doctor of physic' and a surgeon testified as to the poor state of Smith's health, he was sentenced to two years' hard labour. The prisoner responded: 'This sentence is equal to a sentence of death upon me.'[22] It is not known if he survived.

Daniel Eaton was a constant thorn in the Establishment's side and faced no fewer than eight prosecutions between 1793 and 1814. He too was convicted of selling *A Summary of the Duties of Citizenship!*, despite Parkinson giving evidence that Eaton's wife had refused to sell him copies of the work after 'finding that they were deemed unsafe'.[23] Aware of the sentence that had been passed on Smith, Eaton decided to abscond to America, rather than face the inevitable conviction. He remained there for four years.

Paul Lemaitre also spent more time in prison, seized by the authorities, he believed, simply because he had been arrested previously. In 1846, 50 years after the original trial, Lemaitre was still petitioning the Government to formally recognise his innocence and return his papers.[24]

The Gagging Acts had a deleterious effect on the LCS, driving most of its activities underground. Then in February 1797, 1,400 French troops landed unopposed on a lonely part of the north coast of Pembrokeshire, two miles to the west of Fishguard. The war against France had been in progress for four years and the invasion was part of the French Directory's plans to invade the British Isles. The Commander-in-Chief of the French troops was Colonel William Tate, an Irish-American hostile to the British. He expected to be welcomed with open arms by the British whom he believed were repressed and starving. When that did not happen he surrendered three days later to a hastily assembled militia. Although the invasion was easily overcome, it came as a tremendous shock to the British people and Government.[25]

The scare caused by the French invasion was followed in May of that year by two significant naval mutinies in which the London Corresponding Society was believed to have been involved, and the Government once more stepped in. In April 1798, all London Corresponding Society committee members were again arrested, the *Habeas Corpus* Act was again suspended, and the men captured were again detained in prison without trial, some of them for up to three years.[26] A few defiant members continued to meet following these arrests, but in legislation passed on 12 July 1799 it became illegal for members of the London Corresponding Society to meet at all and the organisation gradually petered out. Despite its ultimate defeat, 'The London Corresponding Society did more in the eight or nine years of its existence, to diffuse political knowledge among the people of Great Britain and Ireland than all that had ever been done before,' claimed its founder, Thomas Hardy.[27] That is undoubtedly true.

During those turbulent years of the 1790s Parkinson lost two more of his children. Baby Wakelin, named after Parkinson's great friend and witness at his wedding, was born in October 1792 but survived only a few months. Even more devastating was the death of his eldest son, James John, in the bitterly cold January of 1795, while Parkinson was waiting to hear when Smith and the others would be brought to trial. James John was just a few weeks short of his twelfth birthday and as the eldest son he would have been expected to take over his father's practice when the time came. Already he was learning how to recognise the different medicinal plants and minerals, and how to compound them and turn them into useful remedies. Now all that would have to be learnt by the second son, John William Keys, who was two and a half years younger.

James Parkinson was one of many disillusioned members who left the London Corresponding Society after the treason trials of 1795 and 1796, which revealed the extent to which its membership had been infiltrated by spies and troublemakers. Perhaps consideration for his family also had something to do with his decision to leave, for he had seen many families made destitute when their fathers had been imprisoned. Recognising how close he had come to imprisonment – and even transportation – must have made him think twice about jeopardising his family any further. Another contributing factor may have been the gout which continued to plague him. His last political statement for many years was his 'Letter to Mr Smith' in Smith's pamphlet *Assassination of the King!* It is a remarkable account of a man prepared to stand up for his principles, whatever the consequences.

7

Dangerous sports

Always be careful to avoid any dog which you see running along, looking heavy and lowering, seemingly inattentive to every thing, his eyes looking red and watery and his tail hanging between his legs, lest it should be mad.

James Parkinson, 1800
Dangerous Sports. A tale addressed to children

IN THE MONTHS after leaving the London Corresponding Society, Parkinson found he had time on his hands. He had become used to sitting up late into the night and found it difficult to break the habit. He therefore turned his attention to a number of books he intended to write – and the next few years became the most productive of his life. In 1799 he published two volumes of a large medical work, *Medical Admonitions*, which must have taken several years to research and write. This was followed, in 1800, by *The Hospital Pupil* (advice to medical students), *Dangerous Sports* (advice on how to bring up children), and *The Villager's Friend and Physician* (advice on 'domestic happiness'). And in case that was not enough for one year, he also dashed off a textbook on chemistry. The result of such endeavour meant he was always busy and saw little of his family; all we know about his domestic life is gleaned from the glimpses he gives us in his written works. In *The Villager's Friend*, for example, he states emphatically that a mother's first duty to her child is to see that it 'is not robbed of that food which nature has provided for it in the breast'.[1] All too often, poor women were paid to act as wet nurses at the expense of their own children. Parkinson strongly disapproved of this practice, reproaching mothers who were tempted by such a 'sordid bribe', since he believed that when a mother 'sells the food of her child, she perhaps also sells its life'. So undoubtedly Mary breastfed their children.

Assuming the child survived infancy – almost half did not – the parents' next duty, according to Parkinson, was the maintenance of its health: 'As health is necessary to happiness, your watchfulness over him in this respect is indispensable.' To this end he wrote at some length on childhood diseases and warned against numerous other dangers that they might encounter. Dogs with rabies were a common problem so when a ten-year-old boy came to see him one day having been bitten by a dog, Parkinson was concerned for the child's well-being. He visited the dog's owners but as the spaniel showed no sign of having rabies, and as he was assured by its owners that it was in perfect health, he treated the child for a minor injury and sent him home. Three weeks later he was called to attend the boy, who now had a high fever. On arriving at the house Parkinson discovered 'the dreadful malady [rabies] was established'.* But although Parkinson immediately removed six or seven ounces of blood, until the boy almost fainted, and the assistance of a physician was obtained, he did not improve. Eventually the child was seized with convulsions so violent as to require two men to restrain him, and within a short time the poor boy 'succeeded [to] a state of quiet insensibility, lasting about an hour, terminating in his death'.[2]

* The fever that accompanies rabies is easy to distinguish from other forms of fever due to the patient exhibiting additional well-known symptoms: a difficulty in swallowing and an extreme fear of water (hydrophobia).

This unexpected outcome made Parkinson visit the dog's owner again, only to find the spaniel still in good health. Furthermore, the owner was so confident of its well-being that he allowed it to lick his face for several minutes. In fact this was a highly risky thing to do. Had the man had a sore on his face, he might well have contracted rabies, for we now know it to be a viral infection that can enter the body through any open wound in contact with the saliva of a rabid animal. Parkinson kept an eye on the dog for a further two years as it roamed the neighbourhood, without it ever showing any sign of being rabid. It was a highly unusual case and ultimately he was forced to conclude, against all prevailing beliefs, 'that a dog, not only free from any characteristic mark of hydrophobia, but apparently in the highest state of health, can communicate by its bite, this most dreadful and fatal malady'.

Some while later he was obliged to write to the editors of the *London Medical Repository* in order to justify 'vague reports lately in circulation' that criticised him for his bold assertion that dogs showing none of the symptoms of rabies could nevertheless still transmit the disease.[3] His response to such criticism was, as always, to write a pamphlet arguing his case. In it he relates another case of rabies he had encountered back in 1790.[4] On that occasion he had been called to attend a young woman who had a sore throat and was having difficulty swallowing. On presenting her with a glass of water she manifested the classic signs of hydrophobia, watching him approach with much anxiety and becoming convulsed as the water touched her lips. Parkinson immediately recognised the symptoms and decided he ought to seek a second opinion, so he called in William Blizard, his friend and mentor while studying at the London Hospital.

While waiting for Blizard to arrive, Parkinson managed to get the woman to drink half a glass of water by dripping it from his finger on to her tongue, making sure it did not touch her lips.

He also questioned her as to where the dog was that he normally saw in the house. She said it had been sent away because it had become so troublesome, but when Parkinson asked if that meant it had become snappish and bitten her, she replied quite the contrary, it had become too fond of her. It had constantly jumped on to her lap, licking her hands and arms – which, Parkinson noted, were very chapped. When Blizard arrived the woman was taken to the London Hospital where she died within 48 hours.

It was a perplexing case. The woman was adamant that she had not been bitten by any dog at any point in her life; the only dog she had had any association with had not shown signs of rabies and yet she most obviously had the disease, so how had she acquired it? As Parkinson was still a relatively inexperienced practitioner at that time, he decided that either the disease had arisen spontaneously, or that it had been communicated by the dog in a condition which 'according to the knowledge we then possessed, was supposed to be incompatible with the existence of the disease'. In other words, the dog did have rabies and had infected the woman, but was itself not manifesting the symptoms.

Having now experienced two such cases – that of the young woman and that of the boy – and even though he felt a 'considerable repugnance' in putting forward the facts because they so contravened prevailing wisdom, Parkinson strongly maintained in his pamphlet that a dog apparently free from rabies could still transmit the disease to people through biting them or licking open sores. This was the first written account to recognise that asymptomatic dogs, as well as rabid animals, could transmit the disease.[5] Having observed the evidence, Parkinson would not be pressured into changing his interpretation – a stance he often took, even when faced with considerable opposition.

Another occasion on which he stood his ground occurred shortly after he left the London Corresponding Society.

Parkinson had just been appointed as a Trustee of the Vestry for the Liberty of Hoxton, the body responsible for the upkeep of highways, paving of sidewalks, illumination of streets, supporting the parish church of St Leonards, and maintenance of parish paupers. One of Parkinson's many duties as a trustee was to collect the Poor Rate, a tax levied on property to provide relief for the poor; another was as Secretary of the Sunday School where children received 'the inculcation of moral and religious principles'. He was greatly in favour of poor children learning to read and write at Sunday School, so it is rather surprising that he spoke out against Samuel Whitbread's attempt to introduce a bill into Parliament which aimed to provide them with a more regular education. Whitbread's Bill proposed that children would receive free schooling for two years in classes to be held on weekdays, instead of Sundays. But Parkinson criticised Whitbread's proposal on the grounds that parents could not afford to keep their children at school for two years because their financial contributions to the family income were often essential. Never one to hold back his views, he stated his concerns in a pamphlet, arguing that the Whitbread scheme, admirable though it was, might destroy existing Sunday Schools. It seems that many others agreed with him since Whitbread's proposal failed to gain parliamentary support and the situation remained unchanged for many more years.[6]

The Vestry was also responsible for orphans and illegitimate children born into the parish. When old enough, these children would be bound as apprentices to masters or mistresses in neighbouring parishes. By moving them outside the parish of their birth, responsibility for them was passed on to the receiving parish so they would not become chargeable as paupers within their own parish on reaching adulthood. However, since all parishes adopted the same scheme, the result was a pointless exchange of dependent children; the parish's burden was not diminished, and the children were exiled from any friends or family they had in the area, leaving them with no one to whom they could appeal if they were ill-treated.

Those who were sent to work in factories suffered particular deprivations. Often as young as five or six, they were frequently employed in the textile industry as piecers and scavengers. Piecers had to lean over the spinning machine to repair the broken threads, which put pressure on the joints – in particular the right knee (or the left, if the child were left-handed). This knee was usually the first joint to give way. William Dodd, a child worker since the age of five, became crippled with the swellings in his joints. As an adult, his right wrist measured 'twelve inches round'. When eventually sent to St Thomas's Hospital for treatment, the doctors decided their only option was to amputate his right arm, further hampering his ability to earn a living. Dodd's sister also suffered a terrible injury when her right hand became entangled in machinery. After she became sleepy, having worked all night, 'Four iron teeth of a wheel, three-quarters of an inch broad, and one quarter of an inch thick, [was] forced through her hand . . . and the fifth iron tooth fell upon the thumb, and crushed it to atoms. This accident might have been prevented, if the wheels above referred to had been boxed off, which they might have been for a couple of shillings',[7] but little deference was paid to such health and safety precautions.

Young girls could also suffer physical damage that harmed their ability to have children later in life. A House of Commons Committee on Child Labour reported that because the female pelvis is considerably wider than that of the male, girls were less able to sustain long periods of standing than boys. Until twelve or thirteen years of age the bones are so soft that they will bend if inappropriate pressure is constantly applied. When having to stand for very long periods, the girls' pelvis was prevented from developing properly and there were many instances where, instead of forming an oval aperture, the pelvis formed a triangular one. William Blizard was one of the surgeons who contributed evidence to the House of Commons Committee:

> When they are expecting to become mothers, sometimes because of the development of the bones of the pelvis, there is not actually space for the exit of the child which is within the womb. Under these circumstances, it is often the painful duty of the surgeon to destroy the life of the child in order that he may preserve the more valuable one of the mother. I have seen many instances of this kind, all of which, with one exception, have been those of females who have worked long hours at factories. I believe if horses in this country were put to the same period of labour that factory children are, in a very few years the animal would be almost extinct among us.[8]

Like Blizard, Parkinson was also concerned for the welfare of working children and was moved to act when a child apprenticed in an adjacent parish was murdered by her mistress. The Vestry was responsible for some 70 orphans, most of whom were required to work, but those apprenticed in other parishes were often left to the almost unrestrained caprice of their masters or mistresses, 'no law existing by which the duties of the master are defined, or any inspectors of his conduct appointed'. Parkinson proposed the introduction of a register of poor children seeking employment and advocated measures to monitor their working conditions, suggesting they were 'visited by a committee of the trustees and overseers of the poor twice every year'.[9]

The Vestry appears to have accepted his recommendations, as a panel of inspectors was immediately appointed to make regular visits to the homes of apprentices. They checked that the children were being adequately fed and clothed, that they were not made to work excessive hours, and that they were being trained in work that would enable them to earn a living when the apprenticeship expired. Six months later the first inspections had been completed. The accompanying report illustrates how Parkinson's concerns about ill treatment had been fully justified, with the committee finding itself under the 'painful necessity' of reporting a 'shocking instance of seduction and depravity'. A young girl had been seduced by her master, a married man and father of six children. On discovering that she was pregnant, the master ran away, leaving the young girl to deliver her child in the workhouse; tragically, both mother and child died shortly afterwards.

Parkinson served on the new committee set up to oversee these children and made detailed reports of his visits; these show that he called on no fewer than 72 houses each year, spread over a wide area of the parish. He then established a set of regulations governing the apprenticeship of these children, which included not allowing the children to work on Sundays and making sure that they went to church at least once every week. No child could be apprenticed before the age of twelve and they were not allowed to 'work longer than twelve hours in any one day, and not before six in the morning, nor after nine o'clock in the evening'. Their masters were expected to furnish the children with new clothing on the first day of May each year, and no more than two children were permitted to sleep in a bed. Finally, each child was to be given a copy of the regulations – even though many could not read – and the name of someone in the Vestry to whom they could apply if they were mistreated. Six years later the committee noted that things were much improved: 'Mr Parkinson reported that the Officers, himself and several of the Committee, visited the children apprenticed and found them in general comfortably situated.'[10]

While the introduction of Parkinson's recommendations brought about a marked improvement in the care of destitute children in the local parish, it was another 30 years before laws governing child labour were introduced nationally. Following the testimony of former child workers like William Dodd, changes first came in 1833 when the Factory Act was passed. The Act made it illegal for textile factories to employ children younger than nine, and stated that the working day was to start no earlier than 5.30am and to cease at 8.30pm. A child aged between nine and twelve could not be employed for more than nine hours in any shift, and a young person (thirteen to eighteen) for no more than twelve hours. A succession of laws followed during the mid-nineteenth century, but it was 1901 before the age at which a child could be legally employed was raised to twelve. At a time when society was largely indifferent to pauper children's working conditions, James Parkinson made a significant difference to some of their lives, almost a century before it was a legal requirement to do so.

When it came to managing his own children Parkinson was 'doatingly fond', but perhaps due to having lost three when they were young, he appears to have been excessively worried by the daily risks they faced. Once children had passed the age of five, it was less common for them to die of a childhood disease, so perhaps eleven-year-old James John had died in an accident which subsequently made Parkinson fearful and overly anxious for his surviving children. In 1800, five years after losing James John, Parkinson wrote a book called *Dangerous Sports, a tale addressed to children*, which warned against 'wanton, careless, or mischievous exposure to situations from which alarming injuries so often proceed'. The book is dedicated to parents and schoolmasters, with the quotation on the front page encapsulating its essence: 'Who knows but one of my stories may one day save the life of some child.'[11]

'Benevolence on Crutches' tells the story of lame Old Millson who lived in a cave dug out from the side of a high and craggy mountain, somewhere down in the West Country. One cold and dark December evening, when snow lay on the ground, Millson heard a plaintive cry coming from outside. Snatching up his crutches, he went out to investigate and discovered a young boy who had been thrown from his horse and was lying unconscious on the ground, the blood from his wounds staining the snow. With great difficulty, due to his lameness, Millson managed to carry the boy to his hovel where he nursed him back to health, eventually reuniting him with his distraught parents, who were, of course, eternally grateful.

Such maudlin writing was typical of the time. Sentimental novels in particular featured scenes of distress and tenderness which sought to forge links between author and reader. Authors would portray their characters as having 'fine feelings' so they came across as refined, sensitive and morally upright individuals, while scenes of those in distress aimed to engender 'feelings of a tender heart, the sweetness of compassion, and the duties of humanity'.[12] Henry Mackenzie, author of *The Man of Feeling*, explained that sentimental writing worked by depicting situations recognised by readers, allowing them to feel 'that pleasure which is always experienced by him who unlocks the springs of tenderness and simplicity'.[13] Parkinson was a master of the genre, using it to powerful effect to raise awareness of issues concerning the poor, injured, disabled, sick, and dying, although the style was beginning to look a little old-fashioned by the early 1800s.

During the scene described above, the young boy became so devoted to Millson that his dearest wish was that the old man should attend his birthday party, along with a dozen of the boy's friends. Parkinson then used this contrivance to ensure that every boy attending the party got into some kind of difficulty, so Millson could hold forth about how reckless they had been and how close they had come to death or, at the very least, being 'crippled for life'. The list of 'dangerous sports' during which accidents could happen is extensive, including such bizarre possibilities as falling into a tub of blood kept for making black pudding when playing hide and seek in the dark. One child even drank from a bottle of nitric acid, which so badly burned his mouth and throat 'that he died in the greatest agonies'.

The antidote to such hazardous pursuits, Parkinson recommends, is to sit quietly at home reading a book, learning by rote 'some little geographical table ... the characters of some plant, or the natural history of some animal', or examining the night sky 'bespangled with suns and other worlds'. The microscope was also a source of endless instruction, demonstrating how 'works of art are exceeded by those of nature', and providing far more amusement than 'two or three of those foolish toys which are often destroyed weekly'. And if exercise was required, what better than a game of shuttlecock? 'It is truly curious to see, in this sport, that almost every muscle in the body is called into action and that the whole might of a man may be employed to combat four feathers and a cork,' he muses. We can well imagine Parkinson playing shuttlecock with his children in the park opposite their house.

Despite his deep concern for the well-being of his children, he was nevertheless quite a disciplinarian, as was common at a time when a strict code of etiquette controlled social behaviour, and children who misbehaved could ruin the reputation of their parents. Thus he entreats parents to regulate their infants' passions and teach them to distinguish between right and wrong, lest the child becomes a 'wretched nuisance' which would 'render him odious to all around him'. At the same time, it was important to administer restraint with mercy:

> Contemplate the countenance of the poor child who suffers frequent and severe chastisement; observe every feature contracted by habitual terror. The most innocent action is performed with alarm and dread. In justice to the offending little trembler, before you correct your child, correct your own anger.[14]

The abuse of children was abhorrent to Parkinson and he was one of the first medical practitioners to write openly on the subject. A condition he frequently encountered was 'watery head', and although he pointed out that the symptoms could also be occasioned by a fall on the head, he let parents know in no uncertain terms that he recognised the most common cause of the condition: 'I am sorry to be obliged to add another cause of this malady: severe blows on the head, inflicted in the correction of children.'

Many childhood diseases Parkinson dealt with are familiar to us today, but the symptoms were often confounded with similar symptoms in unrelated conditions. When discussing epilepsy in children, for example, he tells the extraordinary story of a seven-year-old girl who was subject to such outbursts of anger that they frequently ended in fits. Rather at a loss as to what to do for this child, who had not responded to any treatment, he tried a remedy recommended by a friend of the family. Around the time of the full moon, two ounces of blood were taken from the child's arm and a teaspoonful of salt stirred into it. She was then made to drink this disgusting concoction: 'This mixture must be swallowed while still warm. If the patient experiences any return of the fits the bleeding must be repeated and the blood again drunk on the ninth morning, mixed with the salt as before.' Despite considering the remedy to be an old wives' tale, to Parkinson's amazement the treatment was a complete success and 'the child was no longer prone to those violent gusts of passion to which she had hitherto been subject'. It seems he had carried out an early form of aversion therapy, the child quickly realising that in order to avoid a repeat performance of this terrifying and disgusting procedure, it was necessary that she submit to the will of her parents.[15]

Parkinson's advice on how to deal with real epileptic patients during a fit was similar to how we would handle the situation today, but the possible causes of 'this shocking disease' were described as being so numerous and so difficult to detect that he recommends always calling in 'some medical mind' as soon as a fit begins. One cause of epilepsy was considered to be the presence of worms in the intestine, but identifying the existence of these presented many problems, as exemplified by the case of a young woman who, 'enfeebled by a weak state of the stomach and bowels', could not be persuaded that her problems were not caused by worms. Without telling anyone, she purchased a quack remedy, advertised as a safe and speedy cure for worms. After a while, her mouth became very sore and she was salivating so excessively she was forced to call in Parkinson. Unfortunately, although he was able to lessen the salivation, she was so 'reduced by the severity of her sufferings, by the quantity of saliva discharged, by the deprivation of her food, etc, that, although placed in the country, under the care of an attentive and assiduous mother, she soon died'.[16]

The 'remedy' she took was probably Ching's Worm Lozenges, which had recently come on to the market. A letter to the *Medical and Physical Journal* relates similar symptoms seen in a young boy following treatment with these lozenges.[17] His mouth began to ulcerate, his teeth dropped out, he suffered from excessive salivation and within 28 days he too was dead. The lozenges contained a 'panacea of mercury' which was intended to 'raise a salivation', a frequent treatment for problems of the stomach. Unfortunately, the large amount of mercury contained in the lozenges made them extremely poisonous.

In relating the story of the young woman who thought she had worms, Parkinson was issuing a warning against both quacks and self-diagnosis. But despite his disapproval, he was sympathetic to those who could not afford the fees of a professional. He would help them, he decided; help them find a way to cope with the numerous ailments they suffered and died from. He would write a medical work that detailed the symptoms of each disease and tell the public when it was necessary to go to the expense of calling in a physician, and when it was not. As he sat down to write, little did he know that the work was to rival the most popular medical book of its day.

8

A pox in all your houses

By inoculation, the shocking ravages of this most disgusting and alarming disease is prevented.

James Parkinson, 1799
Medical Admonitions

AT 825 PAGES, William Buchan MD's *Domestic Medicine: or, a treatise on the prevention and cure of diseases by regimen and simple medicines* was a hefty tome. First published in 1769 and initially selling for just six shillings, it became the country's medical bible, the standard work of reference for family medicine. Some 80,000 copies of the nineteen English editions were sold during his lifetime, making the author a great deal of money.[1] Buchan offered the general public basic, practical advice on health and cleanliness: 'Were every person . . . to wash before he went into company, or sat down to meat, he would run less hazard either of catching an infection himself or communicating it to others.'[2] But although such common-sense advice had proved invaluable for 30 years, Parkinson now considered *Domestic Medicine* to be rather old-fashioned and that 'some few passages in your once favourite book . . . seem to be likely to be more productive of harm than of good'. Concerned that 'many lives are lost by neglecting to apply sufficiently soon for medical aid and by improper treatment of diseases' (and perhaps with one eye on the amount of money he might make from writing such a book), Parkinson wrote a new work on the practice of domestic medicine and the preservation of health. *Medical Admonitions Addressed to Families* was published first in England in 1799 and then in America in 1803.[3] Printed in two volumes and sold at a combined cost of nine shillings, it was probably too expensive for the poor who were most in need of it.

Placing particular emphasis on how to recognise diseases, the work was intended on the one hand to prevent patients incurring the expense of medical attendance for trifling ailments; on the other hand, because physicians *were* expensive and therefore many left calling them in until it was too late, those caring for the sick risked 'sacrificing a friend, or perhaps a beloved child, by delay or improper interference in some insidious disease'. It was a fine balance.

Clearly there were many benefits in being able to determine the nature of a disease in its earliest stages, which would then help decide whether or not a physician was necessary, thus the book's lengthy subtitle ran: '*Directions for the treatment of the sick on the first appearance of disease; by which its progress may be stopped, and a fatal termination prevented from taking place, through NEGLECT OR IMPROPER INTERFERENCE*'. The work proved highly popular and was an immediate success; a second edition appeared later that year and a third edition the following year. Favourable reviews undoubtedly helped sales:

> *Monthly Mirror*: Such families as are hostile to the frequent visits of Medical Practitioners, will do well to lay out Nine Shillings in the present work.
>
> *Medical and Physical Journal*: We have perused the above work with uncommon satisfaction, with the design and subject, as well as the manner in which it is executed.

Monthly Magazine: Such a work is truly a *desideratum* in English literature, and cannot fail to be productive of the most beneficial effects.

Ladies' Monthly Museum: We think it a valuable accession to every domestic library, and we promise every family much use from the diligent perusal of its contents.

Monthly Visitor: No family should be without it. It is an important and useful Work.

Critical Review: We had ourselves such a Work in view, but are glad to find it in such able hands.

Medical Admonitions dealt with every disease imaginable – more than 80 across the two volumes. It started with a Table of Symptoms, for it was symptoms that enabled the physician to 'steer his course with confidence and safety, and to discover the dangers which threaten'. So, the patient was asked, was blood being discharged through coughing, vomiting, spitting or the fundament (rectum)? If coughing up blood this generally meant consumption, in which case early attention by the physician was essential; vomiting blood was also 'attended with no small degree of danger' and again the physician should be called; but spitting blood without coughing or vomiting, was 'unaccompanied by danger' and so there was no need to go to the expense of calling in the physician. If blood was flowing from the fundament, 'This may in general be concluded to proceed from the Piles'; however, if that was not the case, 'some internal mischief is to be feared'.

The Table of Symptoms was followed by a description of each disease and, where possible, ways of distinguishing one from another. For instance, there were no fewer than seven different types of fever: intermittent fever (also called 'the ague'), inflammatory, slow nervous, putrid malignant (typhus), remitting, hectic and, of course, just fever. How could anyone tell the difference between all these in their early stages? By carefully reading this book where the symptoms of each were described and their effects detailed – and then calling in the physician, since treating any fever required 'the utmost exertions, of the most skilful physician, to prevent a fatal termination'. Dr Buchan, on the other hand, had suggested that such fevers would often disappear of their own accord and therefore nothing need be done to relieve them. Parkinson took great exception to this advice to do nothing, considering that if the patient waited until 'the symptoms are so violent, that his life is brought into danger, the opportunity of obtaining his recovery may be past by' and the patient could die from lack of medical attention.

Although Parkinson recognised that the ague – what we now call malaria – was particularly prevalent in the marshy regions around London where mosquitoes bred in vast numbers, he considered it to be caused by 'effluvia' or 'miasma' – poisonous vapours or mist filled with particles from decomposed matter – rising off the stagnant waters.[4] He had no knowledge of its link with mosquitoes and it was to be another hundred years before the true cause of malaria became clear. To explain why so many people suffered from the ague over and over again (the reason why it was also called intermittent fever), Parkinson thought they had a 'predisposition' for the disease, not realising that once you had the ague, it tended to recur. He believed this predisposition could be induced by 'too spare living, excessive fatigue . . . excessive study, indulgence in crude and watery food and in spirituous liquors, and by preceding diseases, particularly such as have been attended with large evacuations. In a word, by every thing which tends to weaken the system.' The advice for preventing a recurrence was to avoid all those hazards – not an easy task.

Treatment for the ague was 'universally known' to be Peruvian bark, a source of quinine, which is still used today to treat malaria. The reddish-brown, bitter-tasting powdered bark had been introduced into Europe in 1640 and its efficacy as a therapy for the ague was first developed in tests performed on ague patients living in the salt marshes of Essex during the mid-seventeenth century.[5] To hide its bitter taste, the bark was given as an infusion in white wine.

In cases where Parkinson advised that the physician should be called, it was imperative that once the great man's advice had been received, at no time was it to be contravened by 'the timid friends of a patient' who felt the prescribed remedy was too harsh, or any 'presumptuous nurses' who considered they knew better than the physician. Not that Parkinson harboured any delusions about being able to change people's beliefs in 'domestic quackery', as he called their prejudices: 'I entertain no wild expectation of conquering the prejudices of doctresses and of nurses themselves.' Nevertheless, he strongly disapproved of any interference whatsoever in the physician's advice, even if it was done with the best of intentions.

But much worse than those who were ignorant but well-meaning were the backstreet quacks who deliberately sold their potions, liniments and other remedies to the poor and vulnerable, making outrageous claims for their efficacy despite knowing they would do more harm than good. Street vendors sold patent medicines on a 'no cure, no pay' basis while others offered an array of potions, powders and elixirs to those most desperate for relief from pain or discomfort. Parkinson railed against these quacks and was particularly outraged by those who claimed they could cure breast cancer, even though their remedies had been tested by surgeons of the 'first abilities' and were found to be useless. He was outspoken in his condemnation: 'cruel, daring and I may, with the strictest propriety say, murderous quacks, are hourly pretending to cure this disease.' But not only did the quacks delude their unhappy patients into believing in their nostrums, they also tried to persuade them to put aside all ideas of being operated on, which Parkinson knew was the only cure for breast cancer.

And then there were the quacks who tried to pass themselves off as physicians or apothecaries when they had had no training whatsoever. But telling the difference between a real doctor and a quack was not easy, as even Parkinson admitted: 'I am here at a loss to answer you . . . But avoid the man of coarse, bold and assuming manners. Not only in such will you be least likely to find sympathising tenderness; but, from his vulgarity you may conclude, he has not had that education which is necessary for the foundation of true medical knowledge.' And if times were hard and the patient could not afford to pay a physician? Well, then it was necessary to 'make your case known, without reserve, to your rich neighbour, and fear not a repulse – benevolence is the characteristic of Britons',[6] though it seems unlikely that many of the poor in Hoxton had rich neighbours.

Perhaps recognising that his two-volume work might be too expensive for those who needed it most *The Villager's Friend and Physician* was, at just one shilling, aimed at those who could not afford *Medical Admonitions*. There was even a sixpenny version, a poster entitled *The Way to Health* that the labourer could hang by his fireside and refer to when necessary. Published the year after *Medical Admonitions* and subtitled 'a familiar address on the preservation of health and the removal of disease on its first appearance; with cursory observations on the treatment of children, on sobriety, industry etc', *The Villager's Friend* not only addressed the reader's physical well-being, it considered their moral welfare as well. As one reviewer remarked, moral advice was 'the subject on which the humble and laborious classes of society, stand in most need of'. The booklet's paternalistic tone sounds offensively condescending to a modern reader, but when Parkinson was writing on family medicine he was addressing a newly literate working class who readily accepted such patronage. Through hospitals, dispensaries and pamphleteering, Londoners were exposed to a snowstorm of such advice that advocated fresh air, body washing, clean linen and clean living, so a medical man caused no offence when he urged them to be sober and to use common sense in caring for their health.

Parkinson wrote for this new market and *The Villager's Friend* became so popular that four years after publication in London, several American editions appeared. A review in an American magazine praised this work 'from the pen of that worthy philanthropist, Mr. James Parkinson', implying the author was by now well-known in that country. Indeed, the reviewer so admired Parkinson's 'true spirit of moderation and benevolence', he quoted extensively from the publication to illustrate his point.[7]

Hints for the Improvement of Trusses,[8] on the other hand, although also written 'for the use of the labouring poor' and therefore aimed at the same market, was not just an appeal to common sense. Many labourers were frequently subjected to such heavy manual work that hernias were a common condition. The dangers of neglecting them could not be overemphasised and the only hope of checking their progress was to apply pressure by means of a truss. Unfortunately, trusses were expensive and generally out of the reach of those who required them most, largely due to the fact that their designs were patented.

Parkinson found this situation unacceptable, deploring the fact that such a vital surgical appliance could be patented: 'Contrivances by which the conveniences or the luxuries of life are increased may, perhaps, be considered as fair articles of pecuniary speculation', he admits; but, he asks, should items necessary to the preservation of life be monopolised in this way? 'Certainly not.' He went on to castigate those who exploited their fellow human beings in this way and decided to thwart

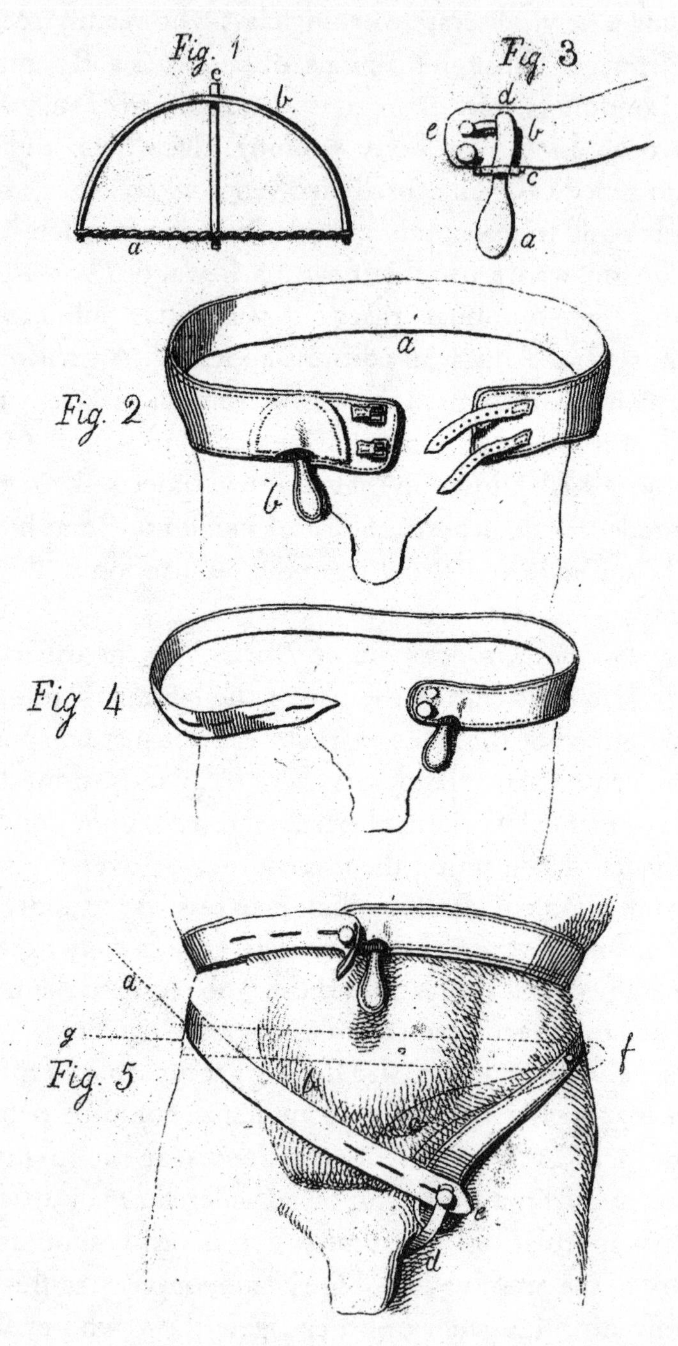

The five stages in making the truss designed by Parkinson.

them by designing his own truss and publishing the results for all to see. His hope was that the patient would be able to make it himself, since it was 'more simple and efficacious than any which has been, as yet, adopted'. The description of how to make the truss was therefore accompanied by a sketch showing the five different stages in its construction. Parkinson must have felt justly rewarded by the complimentary remarks about his philanthropic attitude subsequently made in various monthly reviews.

Exploitation in all its forms was rife in Georgian society as, indeed, it still is today, but there was one disease that was particularly susceptible to domestic quackery. Smallpox was a terrible and terrifying illness which, Parkinson tells us, could leave women so blemished they were unable to get husbands and 'many a handsome face . . . so disfigured as hardly to bear a resemblance to the human form'. In the worst cases it could kill – indeed, 10 per cent of all deaths were due to smallpox, this number rising to 20 per cent in towns and cities where infection spread easily. Among children it accounted for one in three of all deaths.

The term 'small pox' was first used in Britain in the fifteenth century to distinguish it from the 'great pox', or syphilis. It came in two forms – the 'distinct' which was far less serious than the 'confluent', these being so called because the pustules were 'distinct' and separate from each other in the first kind, but merged together in the confluent form. When the scabs flaked off they left ugly scars, the face being the worst-affected area because the distribution of the rash was always more dense there. Blindness could result if the eye became blistered, and limb deformities occurred when the bone or bone marrow became infected.

In the distinct kind, the pustules are sharply raised, typically round and firm to the touch, having the feel of a small bead under the skin. Fluid constantly leaks from them so they slowly deflate and eventually dry up to form scabs. In the confluent form, the fever is greater and the eruption of pustules occurs earlier in the disease. Although the pustules are smaller, initially, than those of the distinct form, they are more dense and, as Parkinson explained, 'the spots assume a crimson colour: they do not rise like the distinct kind but, remaining flat, and running into one another, they very much resemble the measles, during the first days of the eruption. As the eruption proceeds, little vesicles form on the top of the pimples.'

Parkinson described how the head and neck of the patient swell to a considerable degree at the commencement of the 'eruption', the saliva becomes so acrid that it strips the skin from the inside of the mouth, and the pustules become a purplish colour due to the blood dissolved in them 'which, in these cases, escapes at every outlet'. In the worst instances, small purple spots like flea bites appear between the pustules, and blood is discharged with urine and stools. After about sixteen days of this misery, a second fever erupts and gradually increases, during which time the patient becomes excessively anxious and restless. Eventually they subside into a delirious or comatose state and breathing becomes more difficult until eventually 'Suffocation, or violent convulsions at last puts an end to the life and sufferings of the patient'.[9]

Distinct smallpox was treated with fresh, cool air and a light diet with plenty of fluids, but there seems to have been little treatment for confluent smallpox, though Parkinson does offer some advice regarding common practices. Domestic quackery advised that the patient should be put to bed 'under a load of bed-clothes, in a room heated with a large fire, the fresh and pure air being, as much as possible, excluded'; furthermore, the patient was constantly to be supplied with hot drinks. Parkinson spoke out strongly against such an 'absurd practice' and was appalled at the 'ridiculous prejudices' that did not allow a change of body or bed linen 'that soon becomes uncomfortable, and even offensive, from the quantity of putrid matter it has absorbed'. 'The advantages arising from cleanliness of the ... patient, and everything around him, must be so obvious ... that to enlarge on them is unnecessary.' In addition, some poor patients were not allowed to evacuate their bowels for several days in the mistaken belief that the pocks would 'fill quicker' if the bowels were full – once the pocks had filled, it was considered an indication that the disease was on the 'turn'. The practice of breaking the pustules on the face of patients with confluent smallpox was also common in the hope that the scarring would be reduced, but James considered it inhumane, since it caused such pain. However, despite his more modern approach to domestic quackery, if the patient survived an attack of smallpox, Parkinson still considered it necessary to purge them – in case they hadn't suffered enough.

The practice of inoculation against smallpox (known as variolation) was already common and many surgeons had built up lucrative businesses administering it. Parkinson considered 'The advantage derived from this practice is so considerable, as to render it deserving of being considered as one of the most important improvements in medicine'. The method involved the physician making a criss-cross of little cuts in the recipient's arm and then applying to the wound a small amount of pus taken from a pustule on a person with a mild form of smallpox. Dr Buchan describes how he administered it to his own son:

> After giving him two gentle purges, I ordered the nurse to take a bit of thread, which had been previously wet with fresh matter from a pock, and to lay it upon his arm [where it was cut], covering it with a piece of sticking plaister. ... At the usual time, the small pox made their appearance, and were exceedingly favourable.

Unfortunately, the identification of suitably mild strains of the disease was difficult and so deaths from inoculation were frequent when the patient was accidentally given a virulent form. Nevertheless, Parkinson carried out the procedure in an attempt to protect his patients from the ravages of the disease and was critical of Dr Buchan 'and other benevolent physicians' who recommended to nurses and parents that they perform inoculation themselves on the children in their care. Parkinson counselled that it was a task for a medical professional, fearing that the 'neglect of due preparation' and 'the want of proper management during the eruptive fever' would ultimately do more harm than good. Furthermore, if anything went wrong, it would alarm the uninformed, reinforcing their prejudices against the practice which would result in fewer people being inoculated.

The year before *Medical Admonitions* was completed in 1799, Edward Jenner, a Gloucestershire surgeon, published a small book on his theory that by contracting cowpox, a mild disease which milkmaids caught from the udders of cows, the milkmaids were protected from catching smallpox. Jenner recommended vaccinating* people with the much safer cowpox serum, rather than the unpredictable smallpox serum.[10] Parkinson had just heard about this exciting new development as *Medical Admonitions* went to press and in its closing pages he wrote that although eradicating smallpox through vaccination with cowpox was still under investigation, 'The further experiments of Dr. Jenner . . . will, I hope, clearly show how much may be expected from this most important discovery'.[11] Over the following two years Jenner developed techniques for taking serum from the pustules of milkmaids with cowpox and drying it on to threads or glass so that it could be widely transported and used to vaccinate people against smallpox. Given Parkinson's support of inoculation, it is not surprising that he took up vaccination as soon as it was introduced.

* From *vacca* meaning cow.

Despite some people fearing the consequences of receiving material originating from cows – one of God's lowlier creatures – an 'Institution in London for promoting universal vaccination with a view to the extinction of the Small Pox' was established in December 1802.[12] Anyone who 'subscribed one guinea annually or five guineas and upwards at any one payment' could become a Governor,[13] the income providing the Institution with funds to implement a programme of vaccination. Parkinson immediately became a Governor with an annual subscription, and in March 1803 he was one of the first doctors in London to offer his services as a vaccinator. A vaccination station was soon set up in the Hoxton Sunday School House and by November that year, 267 patients had been vaccinated.[14]

Later in 1803 the Institution became the Royal Jennerian Society (RJS), with Jenner as president of its Medical Council, on to which Parkinson was soon co-opted.[15] Given his propensity for writing, he was soon tasked with drafting leaflets to persuade more people to be vaccinated and to drop the practice of inoculation in favour of vaccination. Initially the RJS was prosperous and successful, with more than 12,000 people in London being vaccinated in the first eighteen months. As a consequence, deaths from smallpox, which had averaged more than 2,000 a year in the city during the latter half of the eighteenth century, fell to only 622 in 1804.

Unfortunately, many recognised the financial opportunities available to those who established themselves as vaccinators – including the RJS's Resident Inoculator and Medical Secretary, Dr John Walker. Having quarrelled with Jenner and the Society over his expenses, Walker resigned and set up the rival London Vaccine Institution, which soon became 'patronized by many names of distinction'.[16] This led to a significant drop in revenues for the RJS such that by March 1807 its income was only £450, whereas its outgoings were £800. Several vaccinators claimed expenses for rent and 'stores' which rapidly depleted the Society's funds, although Parkinson did not.

In April 1807, a committee appointed by the RJS Board of Directors to examine the activities of the various vaccination stations in the metropolis reported that 'many of them have been lately abandoned and in almost all of them the practice is greatly declined'. This decline undoubtedly contributed to the major outbreak of smallpox in London later that year. 'The Small Pox prevails fatally at Stoke Newington,' reported the minutes of the RJS committee meeting on 26 October 1807. Stoke Newington lies just north of Hoxton and Parkinson must have been worried for his parish. Two weeks later, the epidemic was gaining such momentum that the RJS committee resolved to hold more frequent meetings and to issue diplomas to assistant inoculators in an attempt to get more people vaccinated. It therefore must have been extremely galling for Parkinson and Jenner to read newspaper articles about the outbreak which referred only to the work done by the London Vaccine Institution, with no mention of the Royal Jennerian Society:

> It is a lamentable fact, that the small-pox is at this time, remarkably prevalent and fatal in the metropolis. . . . We are, however, much gratified in observing that the London Vaccine Institution, are taking every measure in their power, to check the ravages of this dreadful disease.[17]

Despite the efforts of both institutions, 169 deaths occurred in the first two weeks of December and questions were being asked as to what had caused this outbreak. The prevailing belief was that it was due to the Small Pox Hospital continuing the old method of inoculation, 'by which means Natural contagion is necessarily kept up'. There was also concern that the serum used by vaccinators was 'not of the right sort' and that it was being delivered 'by persons not qualified', one of whom was 'a bone setter who was formerly a grocer'. Once again, it seems the quacks were at work. However, the RJS appears not to have acknowledged that the outbreak undoubtedly spread faster than it might otherwise have done because many vaccination stations had closed and fewer people were being vaccinated. Nevertheless, despite all these problems, mortality rates during this outbreak were half what they had been in previous epidemics. In September 1807, the Royal College of Physicians estimated that 100,000 people in the British Isles had been vaccinated since Jenner made his discovery.

By all accounts, Jenner was a difficult individual, writing petulant letters about insufficiently supportive colleagues and fighting acrimonious battles with supporters of inoculation, which only alienated them instead of winning them over. Thus he appears to have been at least partly to blame for the Royal Jennerian Society's short life. Parkinson's commitment to vaccination and his desire to stamp out the practice of inoculation meant he remained loyal to the RJS when many were abandoning the sinking ship. He was one of the last to attend its committee meetings and continued to vaccinate his patients whenever possible. The RJS finally ceased operations in July 1809.

During this period Parkinson and Jenner came to know each other well, for they shared a number of interests. In May 1805, both became founder members of a new medical society that was formed when a group of disgruntled physicians, surgeons and apothecaries broke away from the staid and conservative Medical Society of London, then the most powerful and influential medical society in the country. The Medical and Chirurgical Society of London, as the new society was called, eventually became today's Royal Society of Medicine.[18] Here Jenner and Parkinson evidently discussed medical issues in some detail for, in 1808, shortly before the RJS closed, Parkinson gave Jenner his dissecting microscope. Jenner proudly wrote inside the lid of its box: 'This microscope belonged to and was used by James Parkinson, Surgeon, Hoxton, Author of *Organic Remains*'.

At this time Parkinson had just published the second volume of a groundbreaking new work on fossils, to which Jenner's inscription referred. As it happens, fossils were a passion shared by both men.

James Parkinson's dissecting microscope which he gave to Edward Jenner in 1808.

9

The fossil question

Why are the bones of great fishes, and oysters and corals and various other shells and sea-snails, found on the high tops of mountains . . . in the same way in which they are found in the depths of the sea?

Leonardo da Vinci (1452–1519)
The Notebooks of Leonardo da Vinci

IN THE winter of 1785 while Parkinson was attending John Hunter's evening lectures on surgery in Leicester Square and dreaming of being a surgeon, he would have been aware of the work going on two floors above the lecture theatre, which were to house Hunter's spectacular natural history collection. The museum opened in May 1788, the event being attended by 'a considerable number of the literati in which were included several members of the Royal and Antiquarian Societies, the College of Physicians, and many foreigners of distinction'.[1] The *St James Chronicle* reported on the notable event, and in particular Hunter's collection of skulls:

> There is a regular and continued gradation of these from the most imperfect of the animal, to the most perfect of the human species. The most perfect human skull is the European; the most imperfect, the Negro. The European, the Negro, and the monkey, form a regular series. Mr. Hunter observed that in

placing the Negro above the monkey, great honour is done to him; for although a *man*, he can hardly be called a *brother*. He also remarked that our first parents, Adam and Eve, were indisputably black. This is quite a new idea but Mr. Hunter observed it may be proved without difficulty.[2]

Today such comments shock our politically correct sensibilities, but as Hunter explained in his three-hour 'peripatetic lecture', given as the visitors toured the museum, his collection was no haphazard assortment of curiosities, but a carefully ordered series of human and animal parts arranged expressly to investigate and illustrate fundamental principles about life on Earth.[3] John Hunter was evidently thinking about the evolutionary progression from one species to another more than 70 years before Charles Darwin published *On the Origin of Species* in 1859.

Whether James Parkinson was numbered among the 'literati' present on the opening night is not recorded; however, before making his museum available to the general public, Hunter had allowed selected individuals such as medical practitioners and members of London's learned societies to view the collections in May and October each year, believing that his spectacular displays promoted the 'divine art of medicine'.[4] Parkinson had attended on one of these occasions and what he found there changed his life. Hunter's collection not only contained a vast array of human and animal body parts preserved in spirits, over 1,200 bones, skulls and skeletons in all shapes and sizes, 6,000 pathological preparations showing the effects of injury and disease, and more than 800 dried plants and invertebrates, but it also included nearly 3,000 fossils – 'one of the largest and most select collections of extraneous fossils, [that] can be seen in this country'[5] – which immediately became the focus of Parkinson's attention. Years later he was to recall how, 'From the earliest Moment of viewing the splendid and beautifully illustrative Collection of our revered and celebrated Countryman, John Hunter, Remains of Animals . . . became the Subject of my anxious Investigations.'[6] From that moment on Parkinson began collecting fossils, a passion that was to dominate the rest of his life.

Geology was then a young science, a new science still finding its way in the world, and although the nature of fossils had engaged naturalists for millennia, during the seventeenth century debate had raged as to their origins. Beliefs fell into three main camps. One group argued that they were 'sports of nature', or objects created by God to look like shells and other organisms and placed within the rocks for some inscrutable purpose. A second group claimed that 'seeds' such as fish eggs became lodged in soil or rocks where they grew into 'mock' shells, while the third group unequivocally identified fossils as the remains of once-living organisms.[7] By Parkinson's time the theory that fossils were the remains of animals and plants prevailed, but even though this was accepted, there was still the considerable difficulty of explaining how the relics of sea-dwelling creatures could be found far inland and high above sea level. Just how had fossils become embedded in rocks on the tops of mountains?

In the same year that John Hunter opened his museum, 1788, James Hutton,[8] a Scottish farmer with an interest in how the Earth had formed, published his 'Theory of the Earth' in the *Transactions of the Royal Society of Edinburgh*. In it he propounded the idea that the Earth had been subjected to endless cycles of inundations by the sea, followed by uplift of the land. This eternal cycle of destruction and renewal of the Earth – this 'succession of worlds', as Hutton called it – led him to realise that vast amounts of time were required for these processes to occur; as a result, each 'world' must have been 'of indefinite duration'.

Both Hunter and Parkinson were familiar with such ideas and it is possible they discussed them when Parkinson attended Hunter's classes. But even though they accepted that the Earth had been subjected to cycles of the sea advancing and receding from the land, they both found it difficult to explain how this had happened.[9] Furthermore, while some fossils found in Europe were relatively easy to identify, being almost identical to their modern-day counterparts, others either had no modern analogues at all or, even more worryingly, appeared to resemble exotic species now found only in the tropics. Geological evidence was turning up one conundrum after another and it was hard to make sense of it all. So when the discovery of fossil bones in a cave in England revealed that elephants had lived alongside both hyenas, now found only in the southern hemisphere, *and* giant bears, now known to live only in the northern hemisphere, Parkinson and many others were thoroughly perplexed. When was it, he asked, that elephants and hyenas had 'lived together in our climate, were shaded by forests of palms, and took shelter in caverns along with bears as large as our horses?'[10]

It was all very confusing and further confounded by the fact that beneath all the rationalism and worldliness of the Enlightenment, religion remained at the heart of most people's understanding of the world. Practically everyone had a belief in God and the Bible, even though many did not go to church.[11] And although the emergence of geology may have caused some to question conventional beliefs, most who gave the matter any thought at all believed that fossils had arrived at their present locations having been washed there during Noah's Flood. According to the Bible, God had caused it to rain for 40 days and 40 nights in order to punish humanity for its wickedness. This created a flood of such proportions that the whole world was covered in water to the top of the mountains and all living creatures perished, except for those saved by Noah on his Ark. As the waters receded, the remains of once-living animals were deposited in the mud, which covered even the mountain tops, thereby explaining how fossils came to be found at such heights.

For European geologists, the concept of a 'Deluge', as this monumental flood was called, appeared to be supported by the fact that much of northern Europe was covered by layers of clay and gravel in which fossils were sometimes found, as well as erratic boulders that had evidently been moved hundreds of miles from their original sources. Today we know these features to be the result of the many ice ages that have occurred in Europe, but the concept of ice ages was unknown to Parkinson and his contemporaries, and would not be recognised for another 50 years.

Because most people adhered to a literal interpretation of the Bible, they inevitably tried to fit their understanding of geology into that context, but one major stumbling block was that the Bible decreed the world had been created in six days when, from the geological evidence, this was clearly not the case. It was a stricture that many eighteenth-century geologists struggled with as it became increasingly apparent that huge amounts of time were required to allow for extremely slow geological processes. One way around this dilemma was a theory put forward by the Swiss naturalist Jean-André de Luc, who, following the collapse of his family business in Geneva, had moved to England in 1773 whereupon he was appointed Reader, or intellectual mentor, to Queen Charlotte, wife of George III, a post he held for 44 years until his death in 1817.

De Luc was not at all constrained by a literal interpretation of the biblical chronology. Instead, he interpreted geological history as a sequence of seven vast periods. Each period corresponded to one of the seven days of Creation, and each was separated from the next by a 'revolution' that explained the upheavals in the Earth's crust that had caused the land and the sea to change places. The seventh period represented not a day of rest, but the age of human dominance of the Earth. Compared to the vast tracts of time represented by the other six periods, the present world of the seventh period was but a few thousand years old and the 'revolution' that separated it from the others had been the Deluge.[12] Parkinson read de Luc's theory (written in French) and largely agreed with his interpretation of this conundrum.

Another scholar whose work impressed Parkinson was the French zoologist and comparative anatomist Georges Cuvier, who had become an unwitting beneficiary of the Napoleonic Wars when the astounding natural history collection of the conquered Netherlands reached Paris and was placed at his disposal. The collection included a number of elephant skeletons, which Cuvier carefully examined bone by bone, using detailed anatomical techniques. At the end of his research he was able to demonstrate not only that the living African and Indian elephants were two separate species, but also that the fossil elephant, or mammoth, was anatomically distinct from either. Cuvier continued his research, comparing the bones of living mammals with their fossil counterparts, and demonstrated that in many cases the fossils were completely different species which no longer existed. But it still left many unanswered questions, as Cuvier pointed out:

> All these facts . . . seem to me to prove the existence of a world previous to ours, destroyed by some kind of catastrophe. But what was this primitive Earth? What was this Nature that was not subject to Man's dominion? And what revolution was able to wipe it out, to the point of leaving no trace of it except some half decomposed bones? . . . Why, lastly, does one find no petrified human bone?[13]

This last question in particular was central to the debate about whether or not all fossils had been deposited during the Deluge. If, as the Bible stated, all humanity had drowned in the flood, why were no human fossils ever found? Furthermore, if there were species in the geological past that no longer existed today, what had happened to them? At the time it was widely believed that no species had ever become extinct, but Cuvier's work demonstrated this could not be the case – species clearly came and went over long periods of time. But although Cuvier strongly advocated a theory of extinction, he seems not to have confronted the question of where the new species came from, dismissing any suggestion that the differences might be due to the transformation (or evolution) of one species into another.[14] Nevertheless, such ideas had been gaining ground in London with John Hunter and others, as well as in Paris where Jean-Baptiste Lamarck, a colleague of Cuvier's, was an early proponent of transformation, or transmutation as he called it.

Underlying Lamarck's ideas was the principle that Creation was in a constant state of advancement and that it was an innate quality of nature that organisms constantly 'improved' over successive generations; too slowly to be perceived in a lifetime, but observable in the fossil record. Lamarck believed such advances happened because an animal passed on to its offspring the physiological changes it had undergone in its lifetime, and that those modifications came about in response to its environment and survival needs. The long legs and webbed feet of wading birds, for example, arose because the ancestors of those animals needed to feed on fish. In their attempts to wade in ever deeper water, and yet still keep their feathers dry, wading birds would unconsciously adopt the habit of stretching their legs to their full extent, making them minutely longer in the process. This trait of slightly longer legs would then be passed on to the next generation, which would in turn stretch its legs, until, over many generations, the wading birds' legs became longer and longer. Similarly, Lamarck suggested, the habit of birds spreading their toes in order not to sink into the offshore mud, stretched the skin in between their toes and eventually gave rise to webbed feet.

Lamarck was lecturing on these topics before the end of the eighteenth century, eventually publishing his 'theory of acquired characters' in 1809, the year Charles Darwin was born.[15] But although in the end his ideas gave way to Darwin's theory of natural selection, they set the tone for much of subsequent thinking in evolutionary biology. Indeed, in the fourth edition of *On the Origin of Species*, Darwin, having been criticised for not acknowledging previous ideas that had anticipated his great evolutionary theory, added a section on the historical background. He particularly praised Lamark for drawing people's attention to the fact that change in the biological world was 'the result of law, and not of miraculous interposition'.[16]

Although Cuvier did not believe in transmutation, his pioneering work on fossil mammals progressed the new science of geology significantly, and James Parkinson was one of his earliest advocates. Being a surgeon himself, Parkinson well understood when, scalpel in hand, Cuvier proclaimed that comparative anatomy was an essential tool for understanding geology:

> There is a science that does not appear at first sight to have such close affinities with anatomy; . . . in a word, it is only with the help of anatomy that *geology* can establish in a sure manner several of the facts that serve as its foundations.[17]

Parkinson started collecting fossils shortly after he had seen Hunter's collection in the late 1780s. He began by looking for them 'in the gravel pits in the neighbourhood of the metropolis'[18] where he found numerous pebbles that he believed were fossils, but which he could not identify. He collected a large number, breaking them open and examining them under the microscope, hoping that by frequent examination and comparison he might be able to discover their origins. This eventually he did, determining that they were 'some lost species of alcyonium or sponge'.[19] By such perseverance, observation, examination and investigation he gradually enhanced his knowledge and increased his collection. But as the demands of his profession increased there was less and less time for collecting fossils in the field and he soon moved on to bidding for them at auctions, buying them from dealers, and swapping with other collectors.[20]

As his collection grew, Parkinson found it increasingly difficult to identify and classify many of its rare and beautiful specimens, and as little had been published on fossils in English, he was required to read 'the writings of the learned of Italy, France, and Germany'. Having examined all the works in those languages which the British Museum, 'that noble institution', could supply, and having corresponded and exchanged specimens with some 60 collectors, both in England and abroad, he realised that others must be experiencing similar difficulties in obtaining information on fossils in English. He therefore decided 'to take on himself the task of accomplishing, to the best of his abilities, a work of that description'.[21]

But although collecting minerals had become highly fashionable during the latter part of the eighteenth century, collecting fossils was still an unusual pastime and would not become the nation's passion until the 1830s, some years after Parkinson's death. Furthermore, the word 'geology' was not widely understood because the study of it barely existed; it was still considered part of natural philosophy, which encompassed the study of nature and the physical universe. Parkinson himself would use the word 'geology' only once in all three volumes of his work on fossils. So, aware of the potentially small size of his readership, he feared that 'a dry, strictly scientific work' might not recover the considerable expense involved in the undertaking (he bore the cost of publication himself). He therefore addressed it 'to readers in general', in the hope of widening its appeal.

Even before he began to write, he encountered an almost

insurmountable problem – the lack of a terminology with which to describe this new science. Moreover, the word 'fossil' then meant *anything* dug out of the Earth, including minerals and archaeological artefacts. It was only after Parkinson's publications that the word came to refer specifically to organic remains. We must therefore have some sympathy when he complains of having to write about a science that had no name: 'The word FOSSIL appears to be the only word our language can supply, which is capable of being employed as the term denoting these substances.'[22] Despite this handicap, Parkinson embarked on his task with enthusiasm, determined to stimulate his readers with a passion for the science he loved so much.

To accomplish this work, he must have devoted much of his spare time to the project, working relentlessly after his day's work was done, poring over his fossil collection, and reading extensively in many languages. His grasp of the prevailing geological literature is impressive: the historical chapter alone includes references to 54 books and papers. He spent hours in the British Museum examining their 'numerous valuable specimens'; he consulted the library of Sir Joseph Banks 'that friend of science'; the curators of John Hunter's Museum allowed him 'indulgences while inspecting that admirable collection', and the list of illustrious people mentioned in the Preface who helped him in his enquiries is extensive.[23] But in order to complete his research, there was another subject that required his attention.

As an apothecary, Parkinson needed a good knowledge of chemistry in addition to pharmacy, for the two subjects have been inextricably linked since medieval times. Pharmacopoeias and other apothecary books of the eighteenth century included recipes for hundreds of chemical medicines. Conversely, almost all eighteenth-century chemical textbooks presented numerous recipes for the fabrication of medicines.[24] Parkinson therefore recommended that medical students take at least two courses of chemistry during their studies,[25] but not everyone heeded his advice, often being content to study the subject in far less depth than he considered necessary. To these reluctant students Parkinson offered his *Chemical Pocket-Book*, a textbook on chemistry that summarised the latest discoveries in this rapidly changing science, for new findings were being revealed at such a rate it was difficult to keep up.[26] Breakthroughs were made even while Parkinson was writing, hence at the beginning of the book he follows the French chemist Antoine Lavoisier in his description of heat or 'caloric' as an elementary substance, but later on gives the alternative and more up-to-date view then being advocated by Humphry Davy and Thomas Beddoes, that heat was a phenomenon caused by the motion of minute particles of matter.

As with many of his published works, Parkinson read extensively and then distilled his 'assemblage of chemical facts' into a format readily accessible to students and a wider public. In the process he also improved his own knowledge of the subject. The *Chemical Pocket-Book* appeared in 1800 and an American edition followed in 1802. It was so successful that five editions were published in short succession, each advertised as being 'much improved and comprising all the New Discoveries with which the Science has been enriched since the publication of the last edition'.[27] By 1807 the original 230 pages had expanded to 370, there being so much new information to include.

But sharp-eyed observers who perused the back pages of the second edition, published in 1801, would have seen the following advertisement:

> James Parkinson, being engaged in researches respecting extraneous fossils, with the intention of publication . . . earnestly solicits assistance by such communications, and by such specimens or accurate drawings, as may serve to give value to his

> work. . . . for these he will return grateful acknowledgements, and every proper compensation.[28]

By the following year it seems he had received all the information he was seeking on fossils, and had started writing. Suddenly an attack of gout seized him. He developed a severe pain in his hip which extended down his legs into his ankles and rendered him unable to walk. Not being able to stand meant he could not visit patients, and although his son John had just begun his apprenticeship with him and would have been able to take on some of the more mundane tasks, John was certainly not ready to see patients on his own. Having tried a number of remedies such as calomel (a concoction containing mercury) and opium tablets, considered to cure inflammation, as well as various forms of stimulating and sedative embrocations and sudorifics, none of which had the slightest effect, he resorted to blistering the area over his hip, 'the blistered parts being kept open, for about 10 days or a fortnight each time'.[29] It seems unlikely that any of these treatments had any effect and more probable that the attack abated naturally; either way, it was eight weeks before he was able to walk again.

Over the next two years these attacks increased and worsened until his situation became so distressing he feared he was 'sinking under the infirmities of premature old age' (he wasn't yet 50) and rapidly advancing to a state where he would be unable to run his business. He resolved to find relief and developed a plan of action. Believing that his problems were caused by a 'prevalence of acidity' in his system, he attempted to prevent its accumulation by controlling his diet. His food, he tells us, was normally plain and simple 'such as a humble table has afforded' and although he drank wine it was, he considered, sparingly: 'not exceeding two or three glasses at a time'. But there was no getting away from the fact that within an hour or two of drinking even a single glass of wine, the pain in his hip and hands was much worse. The new diet accordingly eliminated all alcohol:

> For breakfast, tea with bread-and-butter was taken: the dinner consisted of the ordinary provision for the family, but vinegar and pickles of every kind, were carefully avoided. Vegetables of the least acescent [sour] kind were used, and articles of pastry, which had often been indulged in, were employed under some restriction. Wines and fermented liquors of every kind were almost entirely avoided. Bread and milk, or milk pottage, formed the supper.*

To correct any prevailing acidity already existing in his body and neutralise acid generated in his stomach, between eight and sixteen grains of soda† 'in its carbonate state' were taken daily. Leeches were applied to the swollen fingers.

Within two or three days of commencing this regime, the stinging and burning sensation had left his fingers, probably due to the effect of the diet rather than any benefit provided by the leeches. But the use of leeches was so routine he continued to apply them even when, after each application, 'the hand swelled very much, became a deep crimson colour, was very hot, and an almost intolerable degree of itching existed all around the wound which the leeches had made'. On one occasion the swelling and redness caused by the leeches extended beyond the elbow, almost to his shoulder. Interestingly, he did not apply leeches to the swelling of his thumb and forefinger on his right hand – probably because that was the hand he wrote with – but he notes

* Pottage was a soup made from vegetables with added milk.

† A grain was nominally based upon the weight of a single seed of a cereal: today it is equal to 64.8mg.

that these improved nevertheless. Somewhat surprisingly for this acutely observational man, he never seems to have considered that applying leeches did not contribute to the improvements in his hands.

Within six weeks the swelling in his ankles had started to reduce and within two months the pain and tenderness in his feet and ankles was almost entirely gone; the pain in his hip was only troublesome when he had walked some distance. But he was apprehensive that the long-term use of such treatment might induce other ailments, as it had in his father, so he suspended use of the soda for a while and relied on the diet to keep the gout under control. Things continued to improve for a further three months 'during which period he did not drink three glasses of wine', such that the finger joints nearly regained their normal size and all pain was gone except that in his hip – which nevertheless continued to improve. Whenever the gout threatened to return he would resume taking the bicarbonate of soda for a fortnight, whereupon the symptoms disappeared again.[30]

But there was one advantage to this miserable period: being forced to sit down, he had time to write his first volume on fossils – though with the pain and swelling in his hands, it is difficult to imagine how this was achieved. He does mention that his eldest daughter, Emma, hand-coloured many of his fossil illustrations so it seems possible that she helped her father with the writing as well – perhaps he dictated to her. However he managed it, publication day eventually arrived.

10
A sublime and difficult science

Impelled by that eager curiosity, which a view of the remains of a former world must excite in every inquisitive mind, the writer . . . long and earnestly sought for information respecting these wonderful substances, from every source to which he could obtain access.

James Parkinson, 1804
Organic Remains of a Former World

THE FIRST VOLUME of James Parkinson's *Organic Remains of a Former World* was published on 1 June 1804. The dedication is to Wakelin Welch, best man at Parkinson's wedding, and reads simply: 'To Wakelin Welch, Esq, of Lympstone, in Devonshire, this work is dedicated, in testimony of the friendship and respect of James Parkinson'.[1] Welch and Parkinson were close friends and shared an interest in geology, Welch writing a rather eccentric text on the subject entitled *A New Theory of the Earth: in unison with the Mosaic account of Creation: with an appendix on the plurality of inhabited worlds.*[2]

Although ultimately a textbook, *Organic Remains* starts as a series of letters between an 'expert' in geology and his intelligent and intensely curious, but less informed, friend. Parkinson, of course, played both roles. In the first letter the enquiring gentleman is accompanied on a geological expedition by his daughter Emma and an endearing companion, Wilton.[3] In real life Emma Parkinson appears to have been interested in fossils, ultimately inheriting part of her father's collection, so it seems likely she really did accompany him on such journeys. Wilton appears to be an entirely fictional character, but his 'resolute scepticism' with respect to rational explanations about the nature of fossils, and his 'submissive credulity' and willingness to believe the more popular explanations as to what fossils were, provides Parkinson with the opportunity to show his readers the absurdity of maintaining a belief in the folklore surrounding fossils, while simultaneously expounding his knowledge about their true nature. At the same time, Wilton's remarks were 'so full of quaintness and of humour' that they greatly amused his companions and, presumably, Parkinson's readers.

Since Parkinson could never resist an opportunity to finger-wag at his audience, the first letter starts with a remonstration to those who wasted their leisure hours 'not seeking intellectual endeavours'. Like Parkinson himself, the narrator of this letter is never one to waste his time, writing that he 'always allotted a small portion of my time . . . to scientific research, and an enthusiastic admiration of the beauties of nature'. With this in mind, he and his companions set out from London in a chaise to 'visit the most interesting parts of this island'. On arriving just outside Oxford they come across labourers mending the road with large, nearly circular, stones that look like coiled-up snakes. 'Curiosity prompted me to stop the chaise, and to ask the man the name of the stone, and where it came from. "This stone, sir," says he, "is a snake stone, and comes from a pit in yonder field; where there are thousands of them."' They all alight and examine some of the unbroken pieces which, though evidently bearing the form of some strange animal, are undoubtedly formed entirely of stone.[4]

'The London Corresponding Society alarm'd' by James Gillray, 1798.

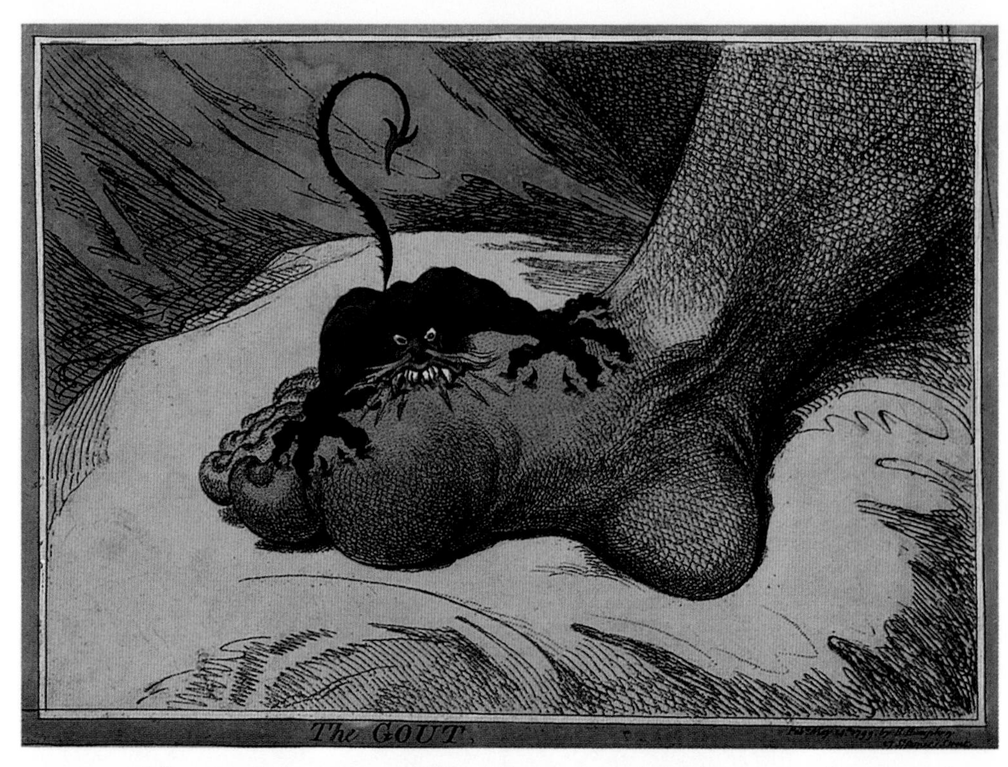

'The Gout' by James Gillray, 1799. James Parkinson suffered badly from gout throughout much of his life.

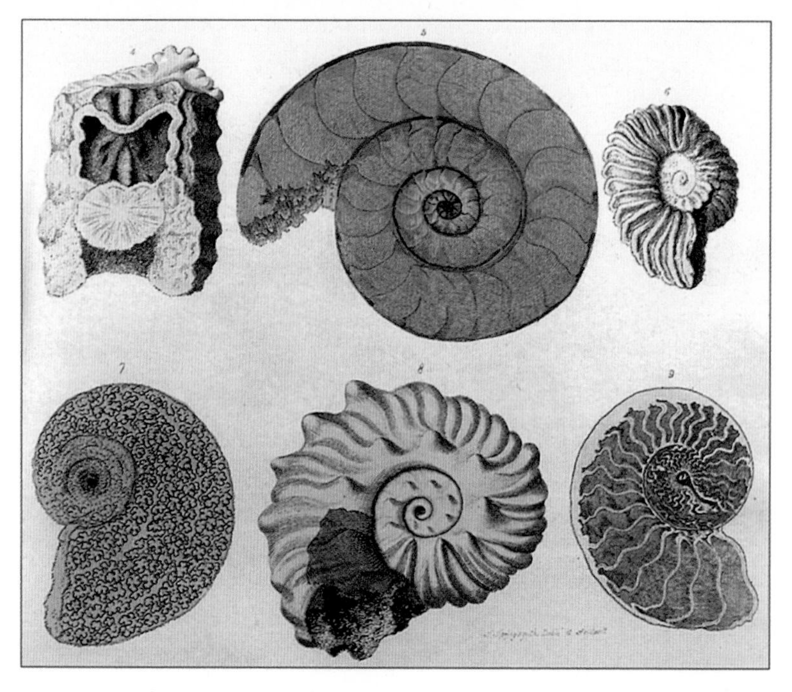

Illustrations of ammonites from *Organic Remains of a Former World*.

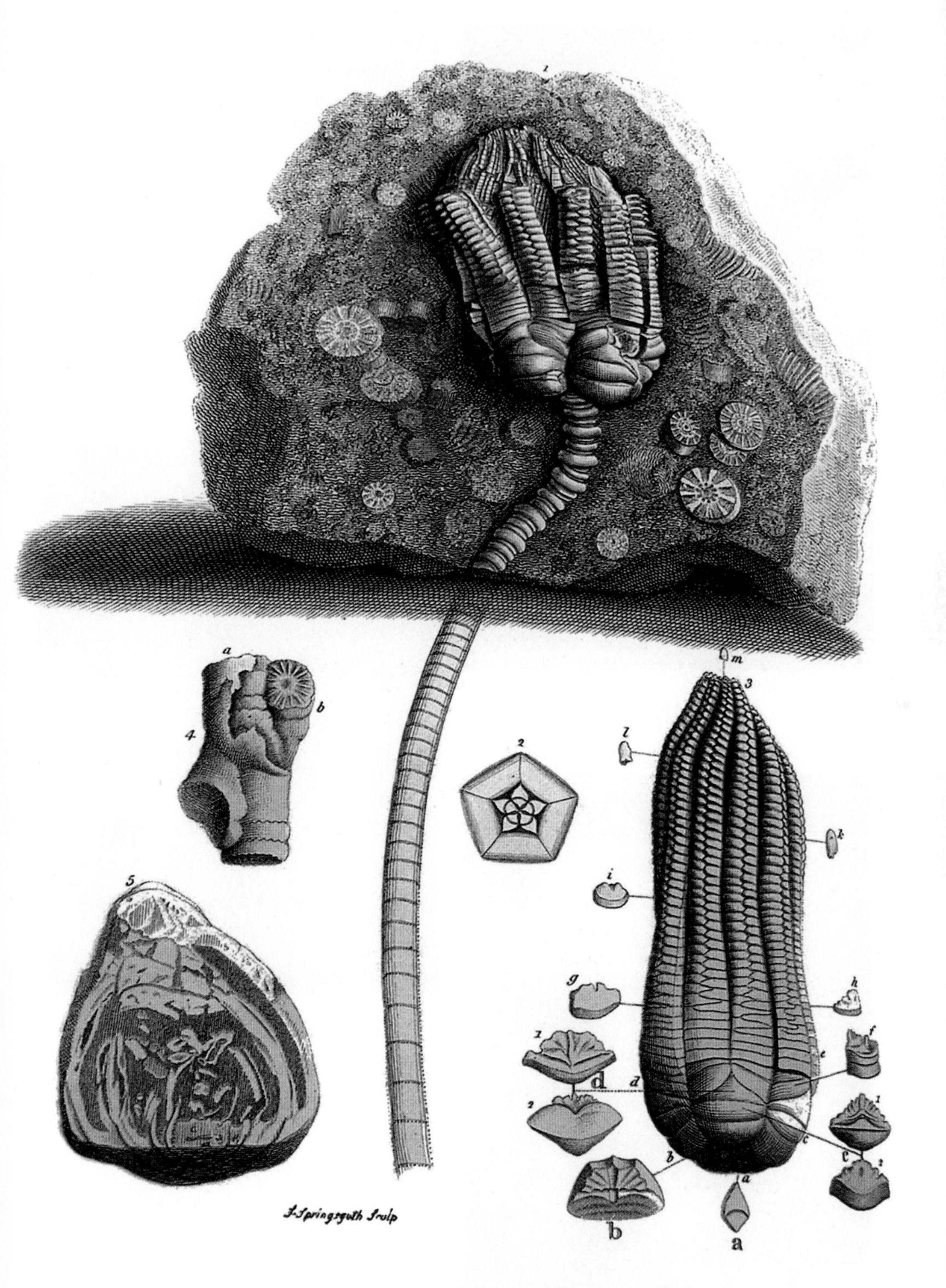

Illustration of a crinoid that Parkinson called
the Stone Lily or Lily encrinite.

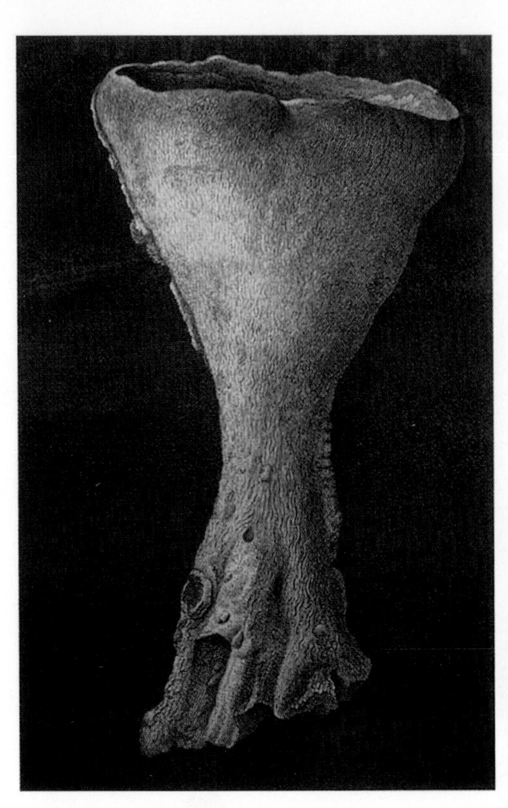

The silicified sponge, *Chenendepora michelinii* Hinde, which illustrates the frontispiece to Parkinson's second volume of *Organic Remains*.

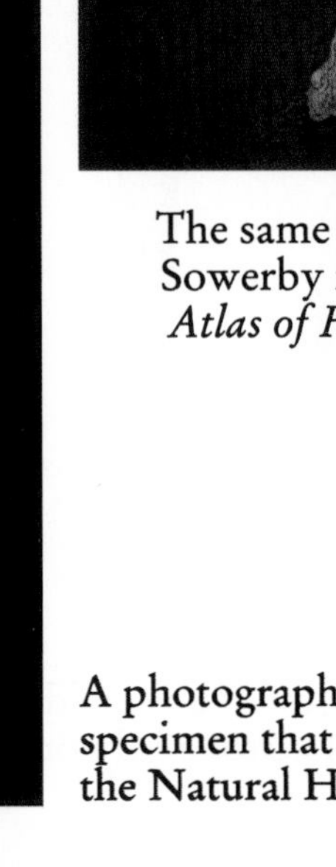

The same image recoloured by Sowerby in Mantell's *Pictorial Atlas of Fossil Remains*, 1850.

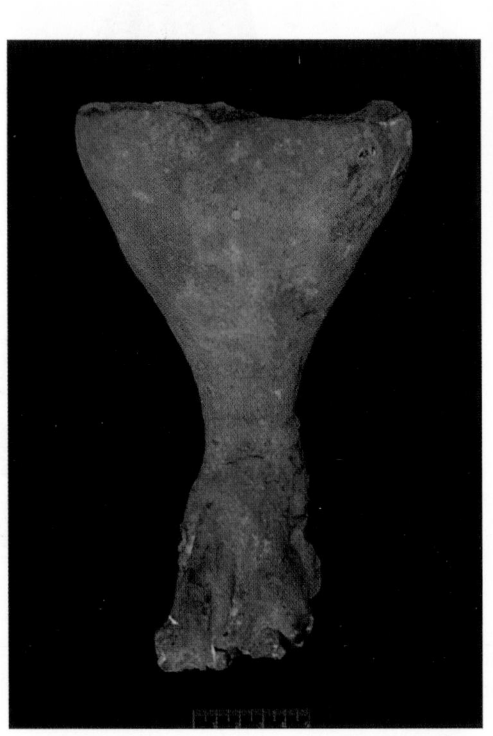

A photograph of the actual specimen that can still be seen in the Natural History Museum.

Hoping to gain further information regarding these interesting objects, the group continue on foot a short distance until they come to a public house 'surrounded by roses and honey suckles in full bloom . . . rendered charming by the wild luxuriancy and profusion', where they stop for something to eat. There they question the landlady about her collection of curiosities displayed on the 'old oaken chimney piece'. Taking up a stone resembling those they had seen in the road, but much smaller, the landlady explains that it is a petrified snake 'with which this part of the country abounds'. According to her, the fossils were originally fairies who 'for their crimes were changed, first into snakes, and then into stones'.

Parkinson undoubtedly knew that 'snake stones' were fossil ammonites, but in order to inform his audience, the letter ends with an entreaty to its recipient – the 'expert' – to 'supply me with a regular, and systematic history of the strangely figured substances; to understand the nature of which, I am impelled by the most eager desire.' In his response, the expert conveys Parkinson's own excitement and wonder about fossils; how so many of them are so different from anything alive today that they 'cannot fail to excite an enthusiastic admiration, and an eager desire to obtain all the information respecting them which can be acquired'.[5] But always in the back of Parkinson's mind as he wrote these letters were questions about how these fossils had been turned to stone and how they came to be where they were found.

Like most of his contemporaries, Parkinson had a profound belief in God and consequently he struggled with the conflict between geology and religion throughout his life. On the one hand he had been brought up to accept the indisputable authority of the Bible, while in the other hand he held a fossil that questioned everything the Bible had taught him. Striving to make sense of the facts that he uncovered about the physical properties of the Earth, he attempted to fit that knowledge into the context of his religious beliefs. Although he tried not to develop theories, it was often difficult to observe rocks and fossils and not wonder about how they had formed, as he explained in a letter to Jean-André de Luc: 'Hypothesis I have ever endeavoured to keep as clear of as I could, but observation of facts will be accompanied by conjecture, oftentimes the foundation of hypothesis.'[6]

In Georgian society, the Earth was generally believed to have been formed in six days and to be less than 6,000 years old.[7] Parkinson's views, however, concurred with those of de Luc; he believed formation of the Earth 'must have been the work of a vast length of time, and must have been effected at several distant periods'.[8] But although he personally might have believed this, he was very concerned not to offend the religious sensibilities of his audience, for he was probably aware of the advice given to John Hunter on submitting a draft of his *Observations and Reflections on Geology* to the country's leading geographer, Major James Rennell. On discussing the preservation of fossils, Hunter had said that they could 'retain some of their form for many thousand centuries'. Rennell advised caution:

> . . . in page 3, you have used the term "many thousand centuries," . . .Now, although I have no quarrel with any opinion relating to the antiquity of the globe, yet there are persons, very numerous and very respectable in every point but their pardonable superstitions, who will dislike any mention of a specific period that ascends beyond 6000 years; I would therefore, with submission, qualify the expression by many thousand "YEARS" instead of "CENTURIES."[9]

It was this same audience of very numerous and very respectable persons with their pardonable superstitions to whom Parkinson addressed his volumes, so he had good reason to be cautious. They were the 'middling sort' of Georgian society; the new, self-confident and affluent middle classes who were in a position to afford and buy his books – 10 guineas for all three volumes of *Organic Remains* – and who also had the education and leisure time to read them. Joseph Priestley considered the middling sort to be 'generally better educated, and have consequently more enlarged minds . . . than those born to great opulence'.[10] But despite their enlightened attitudes, they still considered 6,000 years to be the age of the Earth.

So where had this age come from? Moses certainly had not mentioned any date when he gave his account of Creation in the Bible. In fact, the idea had been conceived only 150 years previously. In 1650 an Irish cleric, Archbishop Ussher, published a scholarly work entitled *Annals of the Old Testament, deduced from the first origins of the world*. Ussher had spent twenty years of his life trying to establish the date of Creation by meticulously examining scholarly works from all over the world and building up a chronology of events. In the first paragraph Ussher wrote: 'In the beginning, God created heaven and earth, which . . . according to this chronology, occurred at the beginning of the night which preceded the 23rd of October in the year 710 of the Julian period'. In the margin of the page, Ussher computed the date in 'Christian time' as being 4004 BC. Twenty-five years later a London bookseller named Thomas Guy (the founder of Guy's Hospital), began selling Bibles with Ussher's chronology printed in the margin. This version became very popular – its success possibly due to engravings of the bare-breasted biblical women it contained, rather than dates printed in the margins – and in 1701 the Church of England adopted Ussher's chronology for use in its official Bible. Over the following decades the figure of 4004 BC for the date of Creation became embedded in the public psyche as having come directly from God. A century later it had become difficult to persuade the public that the new science of geology was providing evidence that challenged

Ussher's interpretation. Despite his trepidations, Parkinson was determined to face this challenge and 'prosecute it with fairness; to shrink from no question on account of its supposed tenderness; and to conceal no conclusion, however repugnant to popular opinion or prejudice'.[11] He was not going to shirk the truth, even if it revealed the Bible to be wrong.

When it came to the question of extinction, Parkinson agreed with Cuvier, believing 'many genera and species . . . which existed before the flood, are now entirely lost'.[12] However, extinction implied such a flaw in God's architecture of the universe that to argue this bordered on heresy. But Parkinson did not flinch. Once presented with the facts, he became steadfast in his views: the 'facts are indubitable' he insisted, despite the 'impropriety of such modes of reasoning'. Rather remarkably for his time, he then extrapolated the concept of extinction to its logical conclusion and recognised that it 'may be taking place even in our days'. In other words, something somewhere in the world was becoming extinct right now.[13] But if some species became extinct, how then did new species arrive on the planet if they had not been placed there when God first created it? From his close attention to the literature, Parkinson must have been aware of the debate around the transmutation of one species into another, but perhaps he considered this a step too far for his audience to assimilate, as he does not discuss it, preferring to put forward the idea of a 'late creation' of new species. Parkinson's God was not content simply to create the world and then abandon it to look after itself, as implied by the Bible; instead He continued to care for the planet, creating new species and placing them on the Earth as others became extinct.

Much more difficult to explain was the thorny question of why no human remains had ever been found in sediments deposited during Noah's Flood. Their discovery would confirm unquestionably that the biblical account was correct but, as Parkinson reasoned, if people had existed at that time, then surely 'their weapons, their various utensils and articles of furniture must necessarily have been frequently discovered among the antediluvian remains'. Since none had ever been found, he was forced to admit 'we are without a proof of the existence of any human beings at the time of the deluge', concluding that 'man had not been created at that period'. But if this was the case, then why, he asked, did the Deluge occur at all, for the Bible taught that its sole purpose had been to destroy mankind? Furthermore, he continued, didn't the existence of a former world imply that the world of today might, in due course, be replaced by another world populated with beings who had even higher powers than ourselves? But such wild speculation was 'dangerous conjecture' and he veered away: 'Why so many beings were created, as it appears, for the purpose of being destroyed – are questions which I presume not to answer.'[14]

The first half a dozen letters of the first volume of *Organic Remains* comprise a general introduction to the subject – the 'pleasures afforded by this science', the 'early existence of such substances', 'opinions respecting the origins of these bodies', and the 'form of the Earth's surface' – after which Parkinson settles down to discuss and examine in some detail the preservation and fossilisation of the 'vegetable kingdom'. In the following 40 letters or so he describes the formation of coal and peat, petrified wood, opals, amber and even fossil flowers and seeds, giving numerous examples of the different types of fossilisation each has been subjected to, and attempting to explain the differences in their modes of preservation, as well as the locations where such fossils had been found and could be seen. It is essentially a compilation of all the prevailing literature on these substances which he has painstakingly reviewed, analysed and reinterpreted for his audience according to his own understanding. It must have been a gargantuan task.

Unfortunately, the first volume did not meet with all the approbation Parkinson must have hoped for. *The British Critic* was generally complimentary, ending its seventeen-page review that largely itemised the volume's contents with the following pleasing remarks:

> We shall, therefore, conclude by observing that a work on the same subject, equally elegant, comprehensive, and impartial, does not exist in English; nor, as far as we know, in any other language. It is written in plain, intelligible, and equal style; such as may, with pleasure be perused by all classes of readers, and is often enlivened by descriptions on topics of considerable entertainment.[15]

But *The Eclectic Review* was quite hostile, and *The Annual Review* even more so. While *The Eclectic Review* considered the scientific content and general sentiments deserved its 'cordial approbation', it particularly disliked Parkinson's letter-writing style, arguing that had the letters been between 'two well-informed mineralogists' this would have initiated a meaningful correspondence between the authors, but letters written by a single individual were lacking in '*oppositional* examination', which politely implied a lack of rigour.[16] In fact the supposed conversation between the enquiring gentleman and the expert ceased after the first couple of letters and continuation of the letter style simply became a contrivance for headings which explained the content of that section. They could equally well be called chapters, so this criticism seems rather harsh.

Arthur Aikin,[17] editor of *The Annual Review*, who had studied chemistry under Joseph Priestley and was himself an amateur geologist, was less inclined to be polite: 'We scarcely know how to characterise the volume before us. It displays extensive reading and a familiar acquaintance with cabinet specimens, at the same time that it betrays an entire unacquaintance with even the rudiments of modern geology . . .' Aikin also disliked the epistolary form, considering it

> a diffuse and vague manner of writing, well calculated indeed to attract the novice, but very unfit to inspire him with that rigorous precision, without which all speculations on geological subjects are no better than idle vagaries of the fancy, better suited to the dreams of the poet than the deductions of the philosopher.

After a couple of pages in this vein Aikin attempted to be conciliatory in his closing remarks by complimenting the coloured plates, but he could not resist one last twist of the knife: 'On the whole, we have derived both pleasure and information from the perusal of this volume, though not as much as we had expected.'[18]

Undaunted, Parkinson immediately started work on the second volume, which he pursued over the next four years, interrupted only by writing a booklet on gout. *Observations on the Nature and Cure of Gout* was published in 1805, while he was following his self-prescribed diet and treatment, and in a period of remission. It relates a number of case histories, including his father's and his own, and details the cure he had found in the hope that his observations might benefit his fellow sufferers. He opens the book with the statement that 'gout is a hereditary disease', notes correctly that men are more subject to the disease than women, and that it 'seldom attacks those who live on a spare diet'. He therefore advises 'that acids of every kind should be used with great moderation; spirituous liquors must never be drunk . . . wines, particularly those of foreign production, and even malt liquors, must be avoided with equal care'. He may not have fully understood the origins of the disease, but careful observation over the years had shown him that 'insobriety, luxury, indolence and voluptuousness' were likely to bring on an attack and should be avoided at all costs.

Attempting to address what, exactly, gout is, he explains that it involves the deposition of a 'concrete saline substance, which sometimes accumulates in considerable quantities, particularly on the joints of the fingers and hands', and recalls that in 1797 Dr William Wollaston had reported to the Royal Society that this 'gouty matter' contained a 'peculiar lithic acid'.[19] Six months later, Dr George Pearson, also in a communication to the Royal Society, had recommended that lithic acid be more accurately termed uric acid.[20] Today gout is known to be caused by an excess of uric acid in the blood, which crystallises and settles in the joint spaces causing swelling, inflammation, stiffness and pain.[21] At the time, no one knew whether uric acid occurred naturally in the blood or whether it owed its origins to problems with the digestive system, brought on by a bad diet and too much liquor. To determine which of these options was correct, Parkinson recommended that someone should undertake a series a series of experiments 'on the blood of the healthy, as well as on that of the gouty' – a remarkably modern approach.

Having stabilised his condition, he was able to pursue his research in more comfort and was busy with the final details for his second volume of *Organic Remains* when late in June 1807 he received a visit from his friend, the chemist William Hasledine Pepys – who had performed experiments on Parkinson's behalf for the first volume of *Organic Remains* – and several of Pepys's colleagues.[22] The group included George Bellas Greenough, the recently elected MP for Gatton who, over the past ten years or so, had developed a great interest in geology. They had come to see Parkinson's collection of fossils and spent the afternoon admiring what Parkinson modestly described as 'a tolerably large and systematic cabinet'. In fact, mention in Samuel Parkes's

Chemical Essays suggests the collection was by now of international significance:

> . . . there are several collections in England which I suspect far surpass those at Vienna. In confirmation I need only refer to Mr. Parkinson's superb collection in Hoxton Square, London. The polished specimens in his cabinet, of the various kinds of wood in a petrified state, are beautiful beyond comparison.[23]

The group must have been impressed with what they saw because a few months later an interesting invitation arrived. Certain gentlemen were meeting at the Freemasons' Tavern on Friday 13 November 1807 to discuss the possibility of forming a Geological Society, and Parkinson was invited to be among their number. Several of these gentlemen had large mineral collections and all of them would have called themselves mineralogists, but latterly their thoughts had turned towards geology. Since Greenough declared that 'Mr Parkinson [was] not merely the best but almost the only fossilist of his day', his inclusion in this new society was essential.[24] Parkinson arrived at the Freemasons' Tavern promptly at five o'clock, along with ten other gentlemen. Two others were unavoidably detained, bringing the total number of founder members to thirteen.[25]

By the end of the evening, a formal decision had been taken to inaugurate a Geological Society, as William Allen recorded in his diary that night:

> 13th. – Dined at the Freemasons' Tavern, about five o'clock, with [Humphry] Davy, Dr. Babington, &c., &c., about eleven in all. – Instituted a Geological Society.[26]

The group set to work with tremendous enthusiasm and the membership grew very quickly. A month or so later all members were sent a booklet entitled *Geological Inquiries*, a handbook on how to observe and record geological information, which requested members to send in details of the geology in their local areas. The opening lines defined 'geology' in case some members were still unsure as to what it meant, concluding, 'Geology in its comprehensive sense is consequently a sublime and difficult science'.[27] And so it is.

Geological Inquiries pointed out that no other country was richer than the British Isles in the mineral and coal deposits on which the country depended, 'and the present time is one in which we are particularly called upon to explore and employ the whole of our native riches and internal resources'. It was a call to arms which few recipients of the booklet could ignore. As the Industrial Revolution surged ahead, the nation's demand for coal and iron was immense, so a Committee of Maps was established and tasked with making a geological map in order to uncover where those treasures lay buried.[28]

Parkinson was immediately elected on to this committee, for when it came to appreciating the importance of fossils to the interpretation of geology, and hence to mapmaking, Parkinson was ahead of many of his contemporaries in following the principles newly discovered by William Smith, now considered 'the Father of English geology'. In the 1790s Smith had been a surveyor employed in the building of canals to take coal and other resources from the mines to wherever they were needed. During excavations for the Somerset Coal Canal in 1794 he noticed that each new horizon of rock they uncovered contained a unique assemblage of fossils that was different to the layers above and below it. And wherever he looked across the country, the same strata contained the same fossils.

By the end of 1795, this revelation had enabled Smith to identify individual strata from the fossils they contained and understand the sequence in which the strata occurred. So, now when looking for coal Smith knew that whenever he revealed a particular rock with a certain combination of fossils in it, he had found the 'signature' rock that lay above a specific coal seam. This knowledge enabled him to advise landowners about where (and, more importantly, where not) to drill for coal, for drilling was an expensive and time-consuming business. Until that time one blue limestone or red sandstone had looked much like any other blue limestone or red sandstone, and so landowners spent vast sums of money thinking they had coal beneath their land when in fact they were looking in completely the wrong part of the sequence.

As Smith travelled around England from job to job he was able to follow the strata from one place to another, gradually building up a geological picture of the country which, from 1799, he recorded on maps. Eventually he became the first person in the world to create a geological map of a whole country. But publishing maps was costly and it took him many years to raise the necessary funds. During those years Smith's ideas about stratigraphy spread through the geological community, although like many new concepts they took some time to become accepted. Parkinson was one of the first to understand the significance of Smith's discovery, which, when combined with his knowledge of fossils, proved invaluable to the Committee of Maps.

Smith's geological map of Britain was eventually published in 1815 and the Geological Society's map on which Parkinson and others had worked (now known as Greenough's map) was published in 1819.[29] Much has been written about the way Greenough and the Society supposedly plagiarised Smith's ideas,[30] but once his methods were known what else could others do but utilise and develop them? That is how science progresses.

The second volume of *Organic Remains* was published in 1808. Although Parkinson still adhered to the letter format throughout all volumes, he had listened to his critics and all mention of Emma and Wilton and their journey around the countryside gets left in volume one. In volume two he gets on with the more serious business of simply reporting the science. Again, the work is illustrated with his beautiful drawings, and it is explained in the Preface that the plates which are not coloured have been taken from other works, while coloured plates have been copied from the fossils themselves (most of which he held in his collection). Each coloured plate in every individual copy would have been coloured by hand – no small undertaking.

This volume dealt with fossil zoophytes, now an obsolete term that then referred to various invertebrate animals, such as sea anemones, corals and sponges that attach themselves to rocks and which superficially resemble plants. Parkinson was particularly interested in the different ways that these organisms had become fossilised and, even though his knowledge of geology must have been quite considerable by this time, he continued to be amazed at the circumstances that can change a living animal into a fossil:

> To what a remote period of past time . . . does this circumstance direct contemplation! A body, differing from any animal substance now known, has been formed . . . in the depths of the ocean of a former world. [It] is now found embedded in a rock, many miles inland, and at a considerable height above the sea.[31]

It must have been difficult to comprehend how such a remarkable event had occurred when there was still no understanding of how mountains formed.

Reviews of volume two were considerably more positive than those of volume one; even Arthur Aikin, still editor of *The Annual Review* and now a fellow founder of the Geological Society, was complimentary:

> The descriptions are full and well compacted, and display throughout that minute accuracy which can only be attained by a long and intimate acquaintance with the general subject of the individual specimens, and are accompanied by figures from original drawings so admirably executed as to leave all former graphical representations of the subject far behind. But besides the reputation of an accurate describer, Mr. Parkinson merits the higher praise of an original discoverer.[32]

The discovery Aikin alluded to was a chemical experiment Parkinson had performed on a fossil coral, a tubipore as he called it, from Derbyshire. Whereas Parkinson's friend William Hasledine Pepys had undertaken the experimentation for volume one, investigating the presence of carbon in fossilised wood,[33] for volume two Parkinson performed his own experiments, illustrating how his confidence as an 'oryctologyst' and man of science was growing.* In doing this experiment he wanted to find out if he could detect any remains of the original animal in the fossil, even though

> I regarded it as almost hopeless to attempt to detect any animal matter in a fossil body which must have existed, in a mineralized state, several thousand years; but as the result, if successful, would prove highly interesting, I resolved on the experiment.

Taking a small piece of the coral fossilised in limestone (which Parkinson called marble), he dissolved the carbonate material by suspending it in dilute hydrochloric acid. Imagine his delight when

* Parkinson used the word 'oryctology' for the study of fossils; the word 'palaeontology' was not coined until 1822. The word 'scientist' was also not used at this time and was first published in 1834. Until then those who pursued science were called 'natural philosophers' or 'men of science'.

> the membranaceous substance appeared, depending [hanging] from the marble in light, flocculant, elastic membranes. Many of these, most unexpectedly, retained a very deep red colour, and appeared in a beautiful and distinct manner, although not absolutely retaining the form of the tubipore.

Parkinson could not have known that the fossil was vastly older than a few thousand years (it was actually more than 300 million years old), for such timescales would not be recognised until the twentieth century.[34] But since it is extremely unlikely that any organic material could have survived all that time, it is difficult to know exactly what he found. One possibility is that it was some kind of chemical precipitate stained red by the presence of iron; another is that the interstices within the tubes had at some point

The fossil tubipore in limestone from Derbyshire (centre) on which Parkinson performed his experiment. Top left, a cross-section of the coral and, top right, the 'animal membrane' which remained after he had dissolved away the 'marble' in which it was embedded.

become filled with elaterite, a hydrocarbon known for its elastic properties. Elaterite can be black or reddish-brown in colour, and is slightly translucent with the texture of India rubber. As it cannot be dissolved in acid it would have emerged from the fossil as the acid dissolved the carbonate material from around it, leaving fronds that approximated the shape of the fossil. In Britain elaterite is a rare substance, but it does occur in the parts of Derbyshire where this particular fossil was found.[35] Aikin, a chemist himself, evidently believed that Parkinson's experiment had solved one of the 'great difficulties attending all theories with regard to the formation of fossils'. And even as late as 1850 Gideon Mantell considered the experiment 'highly important' for having apparently proved that animal membranes when hermetically sealed in solid stone were as indestructible as the rock itself.[36]

Another discovery Parkinson reported in this volume regarded crinoids, marine organisms that even today are the least understood of living echinoderms (starfish, sea urchins, sea cucumbers, etc.), although their skeletal remains are among the most abundant fossils to be found. With a long stalk and cuplike body bearing five, usually branched and commonly featherlike, fronds, at the time it was not clear if they were animals or plants, or even which way up they had lived. It was Parkinson who realised that the fronds were tentacles and not roots, as had previously been supposed. A fellow fossilist enthused about this revelation:

> It is this ingenious discovery and exposure of an error of former writers on these animals, which has greatly contributed to the better understanding of their economy and anatomical details. Mr. PARKINSON's able work on the Organic Remains of a Former World must indeed be considered as the publication of the greatest importance in the study of these remains, and in particular as having given a great impulse in England to their investigation.[37]

Parkinson studied his fossils with a hand lens, in cut and polished sections, and even by shining light through thin slices of them under a microscope, much as we might do today.[38] He also applied his knowledge of chemistry in an attempt to understand how they had formed, but, as we have seen, his demanding practice mostly kept him tied to London, and he had little time to go out into the field to look for specimens himself. In the autumn of 1807 he did visit his geologist friend Joseph Townsend in Bath, and in the summer of 1812 he travelled via Salisbury to Lyme Regis, looking at the geology en route. Undoubtedly he spent time exploring Lyme's famous fossil beach, but whether the tour embarked upon by the protagonist of *Organic Remains* volume one to 'visit the most interesting parts of this island' actually occurred in real life is unknown. A comment written in the margin of that volume by one of his early readers suggests not: 'No one would think this writer had ever wandered further than ye sound of Bow Bells.'[39] Clearly some of his contemporaries felt he lacked field experience.

Over the next three years Parkinson pressed on with his investigations and by 1811 he was ready to publish his third volume. But then, very unexpectedly, a most unpleasant incident occurred that was to seriously threaten his peace of mind and his by then quite considerable reputation.

11
'Tis a mad, mad world in Hoxton

The clanking of chains and the wildness of their cries formed a scene inexpressibly shocking.

Henry Mackenzie, 1771
The Man of Feeling Visits Bedlam

BACK IN 1794, as Parkinson had left Whitehall after being interrogated by the Privy Council, he was given a message entreating him to go at once to Hertfordshire and make enquiries regarding a 'lunatic' patient of his, George Davis. Mrs Davis, George's mother, claimed that the boy had been 'crimped' and removed from her care without authority. Crimping was a method of recruiting people into the army against their will. Undertaken by nefarious persons who were paid by the army for each man they brought in, likely recruits were enticed financially – up to 25 guineas was paid – given free drinks until they were insensible, and then handed to the army before they sobered up. George Davis had been kidnapped by crimpers and was now with the Birmingham Blues, a regiment of the Horse Guards, currently located at Hemel Hempstead, more than 25 miles away.

Despite being drained and exhausted from his ordeal in front of the Privy Council, Parkinson did not hesitate to respond to the appeal for help and immediately set out for Hemel Hempstead

– then a journey of several hours by horse-drawn carriage – accompanied by the boy's mother and one of the keepers from a local madhouse. On arrival, they found George in a hayloft, his face much bruised and one eye almost closed from a blow. They were informed that because he had been 'shamming Abraham' and pretending to be mad, he had been flogged with a whip and beaten with a stick, despite a local doctor declaring George a lunatic as soon as he saw him. This doctor had put George in a straitjacket, telling the quartermaster to remove him 'at his peril'. But even when Parkinson arrived and confirmed the boy's state of mind, and the two doctors had signed a certificate stating him to be a lunatic, the army still would not let him go.

The next day, Parkinson and the distraught mother followed the regiment to St Albans but despite their entreaties they were forced to return to Hoxton without George. Two weeks later Mrs Davis sent this appeal to the *Courier & Evening Gazette*:

> On his arrival at St. Albans, when I found my poor son, who no longer knew his distressed mother, was to be dragged away I knew not whither, I could not restrain my tears and exclamations, which, however were soon drowned by the singing and hootings of the soldiers.
>
> Should any gentleman who possesses the power, be induced, by this statement, to step forward and furnish an unhappy mother with the means of ascertaining whether her son is in existence; . . . they will confer an obligation never to be forgotten, and may be furnished with particulars respecting this melancholy circumstance, by condescending to apply to Margaret Davis, No. 13 New Inn-yard, Shoreditch.[1]

It seems unlikely that anyone replied.

As James Norris Brewer pointed out in 1816, the village of Hoxton had been associated with lunacy for nearly 300 years:

> This hamlet has for many years acquired a melancholy distinction as the retreat of the insane and the city poor. There are three private establishments, of considerable magnitude and respectability, devoted to the former.[2]

The three establishments were Hoxton House, Whitmore House and Holly House. Respectable on the outside they may have been, but on the inside they were institutions of horror. James Parkinson was the visiting surgeon to Holly House, which had opened in 1792 and was run by George and Harriet Burrows.[3] It was considered the more humane of the three madhouses, although today we would consider the conditions appalling. When the first Government report on madhouses was published in 1816, Holly House was reported to be in a bad condition: the pauper men's department was crowded and their privy was in the most disgusting state; part of the yard was underwater and there was an accumulation of rubbish in other areas. In the women's house, soiled blankets were drying before the fire in one room and were 'extremely offensive', and in another room the corpse of a woman who had died the previous day lay on the floor. Furthermore, Mr Burrows was not in when the inspectors called. When questioned as to why the body had not been removed, Mr Burrows said that the patient had been a pauper and although he had reported her death to the parish, they had neglected to take away the body. He was reprimanded for not having set aside a room to which dead bodies could be removed.[4]

Asylums had been the preferred place for parishes to house pauper lunatics in London for more than a century. Holly House had 140 patients, of which 100 were pauper inmates – the remaining 40 being private patients who were paid for by their relatives. The cost to the parish for maintaining pauper lunatics in Holly House was eleven shillings a week, considerably more than the 4s 2d a week they would have cost the parish had they been housed in the workhouse, so Burrows was presumably making a substantial profit and therefore able to afford the improvements required by the Government's report. The following year the situation was much healthier and the inspector was able to report that 'a new room is built as a second story for the pauper men; the yard is clean, and the accumulations of filth, and pools of standing water, which the commission has heretofore reprobated, are removed'. Nevertheless, the inmates still lacked proper warmth and comfort, and many were chained up at night. Despite this, and in comparison with other madhouses, Holly House was considered 'cleaner, better managed, less crowded'. The paupers were kept separate from those who paid and, when questioned, all stated categorically that they had plenty to eat.

The 'Madhouse Act' of 1774 had tried to prevent wrongful confinement by licensing and inspecting private madhouses, setting limits on the number of lunatics confined in any one institution, and by the certification and registration of patients. Nevertheless, many people were still incarcerated when they should not have been. One of Parkinson's duties at Holly House was to assess if patients were insane and decide whether they should be admitted. But, as he pointed out, it was almost impossible to define insanity because the condition could be so readily simulated by cerebral disease or eccentricities of manner. One case in particular challenged him. A gentleman farmer was brought to Holly House by some friends who considered that he was conducting his affairs in a way that would lead him into financial ruin. The farmer was paying his labourers more than neighbouring farmers because, as he explained, he felt their wages to be so low that they were almost starving. 'Am I to be deemed a madman because I will not save myself from ruin by starving a number of my fellow creatures?' he asked Parkinson.[5] Having questioned the farmer for some time regarding his motives and his plans for the future, Parkinson was struck by the logic of his answers and he was no doubt sympathetic to the man's moral stance about paying his workers a living wage. Nevertheless, he was persuaded to sign the certificate of lunacy, although he was seriously concerned as to whether or not he had done the right thing.

Within a short time all his doubts were removed when the man threw himself over the banister of a staircase. Fortunately he was not seriously injured and on being asked what had led him to do such a foolish thing, he said he had been intending to do it for some time but had been waiting for God's consent. That morning he had put a piece of paper on the windowsill to ascertain whether or not God approved of his intention. If the paper blew outwards he was to infer he had permission; if it blew inwards, he did not have permission. Parkinson thereby deduced it had blown outwards. 'No,' the man answered, 'it remained where I placed it, from which I concluded the answer was I might do as I liked, and therefore I threw myself downstairs.' Parkinson reassured himself he had been right to certify the man, correctly surmising that had he not done so and the man had injured or killed himself, he would have been severely criticised for not having confined him.

Attitudes to lunacy at the beginning of the nineteenth century were coloured by moral overtones of 'there but for the grace of God go I'. As the *Universal Magazine* observed, 'Insanity is one of those awful visitations of Providence, which, as we are all exposed [to it], we are all interested in whatever concerns its nature, progress and cure.'[6] Public attention was also being focused on insanity because of the recurrent 'madness' of the King* and the

* King George III experienced a number of 'mad' episodes from the 1780s until the end of his reign. He may have been suffering from a metabolic disorder, porphyria, which manifests short-lived psychiatric symptoms such as anxiety, confusion and hallucinations, although more recently it has been suggested he was experiencing the manic phase of a psychiatric illness such as bipolar disorder.

case of William Norris, a violent and dangerous American lunatic. Norris had made several attempts on the lives of his keepers and so was restrained in irons and confined in isolation in the Bethlehem Hospital for twelve years.* When the press got to hear about this, they accused the hospital of cruelty towards its patients, which raised public awareness about the way lunatics were managed.

This in turn generated interest in the popular pastime of visiting madhouses, then seen as a form of charity: a visit to 'Bedlam' where some of the most famous lunatics resided, cost twopence, which went towards the upkeep of the pauper inmates. In reality, many people went along for the spectacle and as visiting madhouses became more and more fashionable it became increasingly debased with visitors often tormenting and jeering at the inmates.

When Parkinson was appointed as medical attendant to Holly House, all asylums were inspected at least once a year by a committee from the Royal College of Physicians. But although the committee received regular reports of exploitation and neglect, it had no authority to initiate improvements and no powers of punishment, so its reports were generally ignored by those criticised. Furthermore, because the madhouse owners knew when the committee would visit, the house would be thoroughly cleaned and new clothes distributed a week or two beforehand. According to an anonymous account of life in Whitmore House in Hoxton, more commonly known as Warburton's madhouse after the proprietor Thomas Warburton, on the day of the committee's visit Tommy Warburton would eat with the inmates in order that he could say, 'Oh, the patients live so well I frequently dine at their table from choice!' The author of this anonymous

* Soon after establishment of the Bethlehem Hospital in 1247, Londoners abbreviated the name to 'Bethlem', often pronouncing it 'Bedlam', which it eventually became known as.

account, believed to be the poet and journalist John Mitford, claimed that he had conspired with Warburton to admit him to Whitmore House as a lunatic so Mitford could secure his discharge from the Royal Navy.[7] Mitford's nine-month stay provided him with much material to expose the exploitation, neglect and abuse of its confined patients.

William Norris in Bethlem. As a result of the publication of this image, Norris was released from his restraints, but he remained confined in the asylum.

A particularly distressing case was that of Priscilla Wakefield, a well-known children's author who had also written *Reflections on the Present Condition of the Female Sex* in which she called for more educational and occupational opportunities for women. According to Mitford she had been confined for getting drunk and abusing her neighbours, but was 'no more insane than the reader'. When she attempted to write a letter to a friend, presumably complaining of her situation, she was confined in the cellar and strapped down on a bed located between two toilets; these were the only toilets for the whole house of 160 patients, and were emptied just four times a year. Inevitably they overflowed and the floor was permanently wet; the vaulted roof hung with cobwebs and chained up opposite her was a raving lunatic 'whose howlings, execrations and blasphemies continually sounded in her ears; but the hardened hearts of the keepers only laughed at this wretched woman's sufferings, and frequently left her without food for two days together'. She did eventually escape, or was somehow liberated, going on to produce many more books for children, such as *An Introduction to the Natural History and Classification of Insects*, *A Family Tour Through the British Empire* and *Sketches of Human Manners . . . of the inhabitants of different parts of the world*. It seems unlikely she was ever mad.

In the later years of King George III's reign, when he suffered recurrent and increasingly frequent attacks of delirium, it was Warburton who selected the four keepers sent to nurse the King at Windsor Castle. All four were 'fellows whose touch would taint putridity itself, and render it more abominable', claimed Mitford. The King apparently took a great exception to Warburton and would exclaim whenever he saw him, 'Take away that fellow with the long nose – take him away, away, away.' According to Mitford, Warburton had a 'proboscis three inches long'.[8]

Some years later when the visiting surgeon to Whitmore House, John Rogers, was dismissed by Warburton for being 'too humane', Rogers published a pamphlet entitled *A Statement of the Cruelties, Abuses and Frauds Which Are Practised in Mad-houses.*[9] This publication contributed to the report compiled by the Select Committee on Private Madhouses, to which Rogers was called to give evidence. He was particularly concerned that many sane people were incarcerated in such places; the practice of rewarding those who recommended patients, plus the fact that only one doctor was required to sign the certificate of insanity, left the system open to corruption and abuse. As money was the only real requirement for the admission of a patient, public concern increased as stories emerged regarding the ease with which patients could be detained on the whim of a disgruntled husband or by unscrupulous relatives who wished to get their hands on an inheritance in a hurry. One notorious case involved Parkinson.

Mary Daintree's husband had died a few months previously and she apparently blamed herself for his death. Her neighbours complained of her wandering the streets at night calling out to imaginary voices and tying herself up with a cord wrapped many times around her body or arm, saying she had been ordered to wear it and would not allow its removal. Sometimes she fancied she heard voices coming down the chimney, calling her reproachful names and accusing her of killing her husband. Eventually she became so violent and dangerous that it was suggested she should be removed to a private madhouse. She was duly taken to Holly House by her nephew and his wife, Mr and Mrs Benjamin Elliott, with whom she lived. They claimed to be concerned for her safety, fearing her to be suicidal as they had found a knife concealed in her clothing.

The following morning Parkinson was asked to examine Mary, but on finding her answers to be rational – although her manner rather eccentric – he declined to sign the certificate until he had gained more information respecting the nature of her insanity. He went to the house where she had been living with Mr and Mrs Elliott and told them he needed more evidence of her insanity before he could commit her. There he was informed that a lodger and Mrs Daintree's sixteen-year-old son, who also lived in the house, could provide such evidence. The lodger declared that she was convinced of Mary's insanity, having frequently heard her conversing with her voices, and the son stated that he too thought his mother insane, saying 'he had never seen her any more mad than she was then'. Parkinson returned to examine Mary for a second time and after another conversation with her he 'was fully convinced of her being a lunatic', and signed the certificate.[10]

Mary Daintree maintained all along that she was sane and railed against her enforced confinement. One day, after having been confined in Holly House nearly four months, she recognised an acquaintance passing by on the street and called out to her through the window. Mary explained her situation and through her friend's efforts was eventually liberated from her 'abode of anguish'. Almost three years later, Parkinson was subpoenaed to attend the Middlesex sessions as Mrs Daintree had brought an action against her nephew and his wife for having caused her to be committed to a madhouse when sane.

At the trial, Mary Daintree claimed that she had been seized while drinking tea in her rooms by her nephew, Benjamin Elliott, and two female keepers from Holly House, one of whom 'grasped her by the hair of her head, and pulled her down backwards, whilst the other, aided by Elliot, dragged her forcibly down two pair of stairs, put her into a coach, and carried her to the madhouse'. There she was put into a straitjacket until the next day when Parkinson interviewed her. In her evidence she related several acts of fraud, as well as cruelty, that had been perpetrated on her by the Elliotts: 'they took her property from her, and the prisoner – Elliott – made himself master of her houses, furniture and property and possessed himself of her rents and monies wherever he could find them'.[11]

Mary argued that Parkinson had been duped by her nephew who wanted her locked away so that he could get his hands on her money – a common way to remove relatives who were either wealthy or a burden on the family. Parkinson gave evidence for the defence and explained his side of the story, but a dramatic moment occurred when Mary Daintree's son denied ever having been questioned by Parkinson. Unfortunately, when Parkinson was asked if he was certain that the person in court was the son he had questioned three years earlier, he was unable to confirm that it was. He had only seen him once for a very short time and he could not swear that the young man now standing before him was the same boy he had spoken to previously. It seems unlikely that the son would have forgotten the occasion so it is possible he had been primed by his mother to lie in court. On the other hand, if, at the time of Parkinson's enquiries, the Elliotts had produced a boy they claimed to be the son but who was actually someone else, then the real son genuinely would not have remembered the event.

In his summing up, the judge, Mr Alley, emphasised 'the enormity of this wicked and profligate case', the like of which he had never seen before in a court of justice. Elliott was found guilty of illegally conspiring to deprive Mary Daintree of her liberty and sent to Bridewell Prison for six months. Although it was not explicitly stated, the judge inferred that Parkinson had been negligent in his duties and had committed Mary Daintree without proper examination as to her insanity, relying too heavily on the testimony of those who had a vested interest in having her committed.

The case was widely reported in the papers and Parkinson was heavily criticised. *The Morning Chronicle* said that the testimony of the witnesses, which included Parkinson, was 'extremely vague and contradictory'. *The Times* said that Parkinson's examination of Mrs Daintree before confining her was 'minute and circumspect', while *The Hull Packet* said that when Parkinson went to the neighbourhood where Mrs Daintree had lived and enquired as to her sanity, 'he learned that she had acted strangely, and upon this information he signed a certificate that she was mad!'[12] But it was the editorial in *The Statesman* that was the most damning. It questioned why, having initially thought Mary Daintree to be sane, did Parkinson then change his mind?

> How did Mr. Parkinson come to this conclusion? He did not *himself* think the lady insane, but he believes her to be so, on the representation of relatives, whom he might have suspected to be interested in her confinement! And is the deprivation of people's liberty and peace of mind, is the infliction of torture upon them, to be a matter put at the discretion of a gentleman who could argue as Mr. Parkinson did in this case?[13]

Two weeks later, *The Statesman* published a letter from someone who considered Parkinson's evidence had been '*very extraordinary*':

> I would like to ask Mr. Parkinson who made him *Visiting Surgeon* of that Madhouse; by whom was he appointed; by whom can he be removed; and whether he has not a share, or some other interest in the profits of that house; or of some other of that description in his neighbourhood?[14]

Such grave allegations seriously threatened Parkinson's respectability and reputation. Nevertheless, he was not inclined to challenge them because, he believed, 'no one who knew me would give credit to my having committed myself, by conduct so reprehensible as that which had been imputed to me'. In reality he was extremely busy finalising the third volume of *Organic Remains*, which was just about to be published, and probably did not want to waste time having to justify himself. Eventually he was prevailed upon by friends to defend his actions in print and accordingly wrote his version of events in a pamphlet entitled *Observations on the Act for Regulating Madhouses.*[15] In this he continued to advocate his belief in Elliott's innocence and that Mary Daintree had been a 'dangerous lunatic' at the time he had admitted her to the asylum.

At the end of *Observations*, which appeared in print early in January 1811, Parkinson made a number of recommendations as to how the various Acts for regulating madhouses should be reformed. He emphasised the fact that a definition of insanity always presented difficulties – citing a case where the arresting policeman had been put in the straitjacket in mistake for the patient – and recommended that two doctors be required to examine a patient and sign the certificate before a person could be committed, rather than only one as was currently the case. Despite having been an apothecary himself, he argued that only physicians and surgeons should be allowed to sign confinement orders, since 'there is hardly a neighbourhood which is not infested with some ignorant and illiterate being who . . . seeks a living by putting the lives of his neighbours at hazard'.

He further recommended that the reception order confining the patient should be signed twelve or even 24 hours after admission, to give time for further assessment. It should then be subject to immediate review as soon as the patient's condition changed for the better. In cases where a patient's sanity was difficult to assess, he recommended a physician should be called in to arbitrate, and a commissioner sent for if a decision still could not be made. Where testimonies from relations of the patient were needed to determine a patient's insanity, he proposed that the evidence of relatives be sworn on oath, and if the safety of the public was at risk, the Justice of the Peace administering the oath might add his signature to the certificate, in order to give it greater weight. Finally, Parkinson suggested that legal protection should be given not only to the patients but to the keepers, doctors and relatives as well. These were radical proposals, but it needed someone of Parkinson's standing to speak out if anything was ever going to be done to improve the situation.

Although his *Observations* were noted by the Royal College of Physicians who, later that year, advised the Government that the 1774 Madhouse Act needed revising, the Government did nothing until John Rogers published his vitriolic pamphlet some four years later, which described in detail the brutality of the keepers and the degrading regime at Whitmore House and Bethnal House, another institution run by Thomas Warburton.[16] There was a public outcry and a Select Committee was set up later that year by the House of Commons to make recommendations for the better regulation of madhouses. When published, the report carried a surprisingly lurid subheading for a Government publication, which was clearly designed to attract a wide readership: 'This new and very interesting report on the state of madhouses contains numerous cases and some *SINGULAR & SHOCKING DETAILS*'.[17]

A number of monthly publications reviewed Parkinson's *Observations* and applauded him for his desire to vindicate himself and correct the allegations made in the newspapers.[18] *The Monthly Review*, in particular, considered that the attack on his character had been unjust and that he had defended himself 'with temper and moderation'. These publications unanimously agreed that his observations on the prevailing Acts were 'sensible and judicious' and sympathised with the difficulties involved in the certification of patients. In the end, it seems that Parkinson, 'well known to our readers, as the author of some respectable and useful medical works',[19] and his reputation, were not seriously damaged by this episode, but it was another 30 years before most of his suggestions were incorporated into the 1845

Act on lunacy. As with his recommendations for regulating child labour, he was decades ahead of his time.

With this unpleasant business out of the way, Parkinson was able to get back to more interesting matters. The young Geological Society, now five years old, was flourishing. Geology had become an increasingly popular subject for respectable gentlemen to pursue in their free time, and the appeal to members in *Geological Inquiries* to send in facts about their local geology had resulted in a spectacular amount of maps, rocks and fossils pouring into the Society's premises; they had already moved to larger rooms to accommodate it all. As this information accumulated it soon became evident that the Society needed to 'adopt some measures which shall tend to remove the confusion which now prevails' regarding the naming of rocks.[20] Because, as Parkinson had pointed out in the first volume of *Organic Remains*, there was no standardised nomenclature with which to discuss this new science; people in every corner of Britain were calling the same rock by a different name or, even worse, the same name as a completely different kind of rock from another part of the country. It led to many misunderstandings and difficulties when trying to determine the order in which the rocks had been laid down – an understanding of which was crucial when trying to locate mineral resources.[21] A Committee of Nomenclature was therefore appointed to deal with this problem, and a Committee of Fossils, to which Parkinson became the Secretary, soon followed. In recognition of his expertise on the subject, this committee agreed to 'arrange all the organic remains in the possession of the Society, according to the System adopted in Mr Parkinson's work, and to meet every Monday until their objects should be accomplished'.[22] One of the other committee members was Dr Peter Roget, today renowned for his famous thesaurus.[23]

Throughout its early years the Society met every month and members would read papers to each other on their geological activities and discoveries. As time went by it was felt that these papers should be recorded, so the young Secretary, James Laird, and his assistant, Leonard Horner, were tasked with producing a volume of *Transactions*. This eventually saw the light of day in August 1811, as evidenced by a letter from Horner to Greenough, the Society's president:

> My Dear Greenough
> At length may the hour when Young Hopeful [the *Transactions*] will be delivered into the world be declared. Violent throes indicate his approach and in another week I hope my labour pains will cease . . . for they have been violent and incessant. I trust no accident will happen to it as it issues from the womb, for the whole responsibility now rests with me, my brother Doctor [Laird] having gone to visit his Mother at Weymouth.[24]

James Parkinson was one of the twenty or so contributors to this first volume. In his paper *Observations on some of the Strata in the Neighbourhood of London* he opens with a typical finger-wag to his fellow geologists, complaining that they still did not understand the true significance of fossils and how they could be used to elucidate geology, going on to demonstrate a case in point.[25] Parkinson explained how he had examined the rocks in the London Basin and compared them with those in the Paris Basin,* according to the principles laid down by William Smith

* The London Basin is a geological term for an elongated, roughly triangular sedimentary basin approximately 250km long which underlies London and a large area of south-east England, south-eastern East Anglia and the adjacent North Sea. The Paris Basin occupies the centre of the northern half of France, excluding eastern France.

(the significance of which was still lost on most members of the Society). By using fossils to correlate the rocks within the two basins, he was able to show that although small local differences occurred at the two locations, the main rock types were identical in both basins. From this he concluded that 'a former ocean' had once covered both countries – a remarkable deduction for its time. Not only did Parkinson illustrate how Smith's principles could be used to correlate one rock with another rock in two different countries, his research also provided definitive proof that in the distant past the land and the sea had changed places, and that the locations of these two great cities had once lain deep beneath the same ocean at the same time. As one reviewer of the *Transactions* pointed out: 'Mr. Parkinson's contribution opened the gate which led into the new field of stratigraphical geology, a practical application of the science of oryctology [palaeontology]'.[26]

But in addition to the publication of this important paper and the conclusion of the traumatic Mary Daintree affair, 1811 was significant for several other reasons. It was the year Parkinson's mother died, the year his son John married, and the year the third and final volume of *Organic Remains* was published.[27] This volume covered 'the fossil starfish, echini, shells, insects, amphibia, mammals, etc'; and Parkinson had minutely examined every one of these fossils that had come his way and described it in great detail. Now he was collecting specimens not just because they were wondrous and beautiful, but because even small, broken pieces told him something of importance. He reported in detail on the 'clear and so comprehensive' arrangement of fossil shells by Lamarck, many of which were previously unclassified, and on the work by 'the justly celebrated Cuvier' whose rich cabinet of fossils 'dragged to the National Museum [Paris] from different parts of the Continent' placed before him 'a rich harvest which he has most carefully reaped'. Apologising for the 'frequency

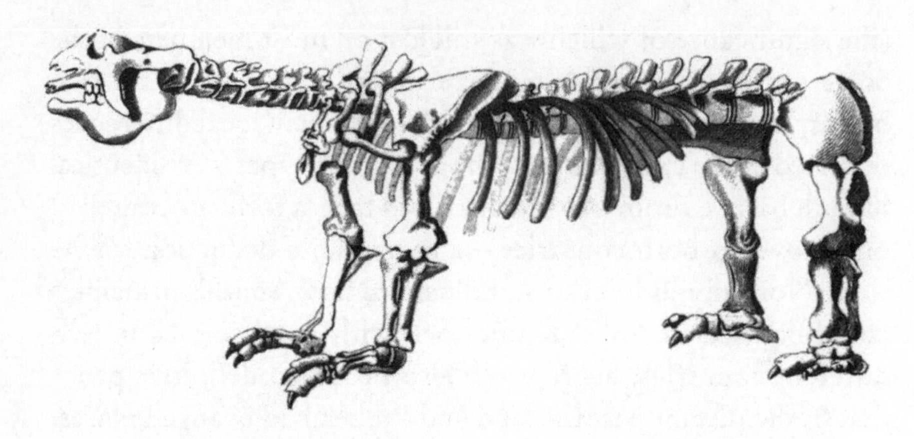

The fossilised skeleton of the extinct megatherium (giant beast) was seven feet high and nine feet long. Cuvier identified it as being related to ground sloths.

with which these invaluable labours are referred to', he explains that to have omitted more of Cuvier's work than he had 'could not have been done without injurious mutilation'.

The 31 'letters' of this volume demonstrate a serious researcher at work who has little time to explain unfamiliar terminology to the uninitiated. The book is now aimed at those who, like Parkinson himself, simply wanted a scientific description and explanation of the fossil under consideration. Only the final letter, 'Fossils considered in connection with the strata in which they are contained', makes any attempt to draw conclusions about the bigger picture. Here he lists the order of strata of the British Isles, as determined by William Smith and published by John Farey,* stating in which strata the fossils described in his volumes could be found. This was in order to show 'how

* John Farey (1766–1826) had worked with William Smith in the early 1800s and was so impressed with Smith's new ideas that he became his 'bulldog'. Farey championed Smith's discoveries at every opportunity in much the same way Thomas Huxley, Charles Darwin's bulldog, was to champion Darwin's work, some 50 years later.

beneficial our enquiries may prove when thus connected' – in other words, how important it was to locate the fossils in their correct positions relative to each other so the order of the strata could be determined.

Smith's now famous geology map of Britain, which would enable everyone to visualise these relationships much more readily, would not be published for another four years. In the meantime the likes of Parkinson and Farey continued their pioneering investigations. Parkinson remarked on how so many strata, particularly those in which coal was found, seemed to have been deeply buried at one time, and marvelled at Farey's estimate that some three miles of strata had been removed from parts of Derbyshire in order to reveal these coals, now found at the surface.[28] Furthermore, 'strata, which appear once to have been continuous, had been broken . . . and dislocated by some tremendous power which has acted with considerable violence on this planet'. No one had any idea at the time how any of this could have happened. Somewhat surprisingly, Parkinson concludes this volume with a rather clumsy attempt to correlate the development of life as illustrated by fossils with the order in which, according to the scriptural account, Creation had been accomplished. By ignoring inconveniences such as the fact that no fossil birds had ever been found,* although they were supposed to have been put on the Earth following the creation of fish and sea monsters, he manages to find 'a pleasing, and perhaps unexpected accordance' between the biblical account and the fossil record. One is left feeling a little disappointed that he still felt it necessary to appease his audience in this way.

Nevertheless, with his three volumes of *Organic Remains*, which described more than 700 fossils, illustrated in 54 colour

* Archaeopteryx, considered to be the first fossil bird, was not discovered until 1860.

plates, James Parkinson had placed British palaeontology on the scientific map at a time when the study of fossils had hardly begun, and when few works on fossils were available in English. One historian has called publication of *Organic Remains* 'the outstanding event in the history of our scientific knowledge of British fossils'.[29] Despite this, Parkinson's hugely important contribution to geological knowledge has been somewhat overlooked, perhaps due in part to the rather convoluted eighteenth-century language in which he wrote, which can be difficult to read. Indeed, even at the time his literary style was considered awkward, as *The Monthly Review* rather harshly pointed out:

> To Mr Parkinson's labours, we cheerfully accord the praise that is due to ingenuity, diligence, and perseverance; and we may be permitted to express a reasonable expectation that, in virtue of his substantial services, the mere geologist will generously overlook a numerous list of literal errors, much clumsiness of style, and a frequent contempt of the rules of grammar.[30]

The Monthly Magazine, however, seemed not to notice these errors:

> The public are deeply indebted to the industry and research of Mr Parkinson, for the investigation of a subject that has never ceased to excite curiosity, but about which only vague ideas and much perplexity has hitherto existed. The present volume closes Mr Parkinson's enquiries, and is not the least interesting of the series because it exhibits on many points his general deductions . . . Mr Parkinson has enriched his work with the most beautiful collection of engravings, colored after nature, that we remember to have seen in any book on these subjects.[31]

Parkinson's prodigious reading in many languages, coupled with his acute observational powers, enabled him to present the latest developments in geology to an audience who had never before seen such exotic and terrifying images of fossils, the 'relicts of a pre-existing state'. As a consequence, his works became extraordinarily popular. Geology was just moving into its golden age and it was largely due to James Parkinson that collecting fossils became the nation's passion during the 1830s. But he also helped move it on from the province of the collectors into the realm of real science, insisting, as he so often did, that fossils could tell us about the formation of the Earth in a way that nothing else could.

For the Romantic poets too, with their fondness for Nature – a wilderness both beautiful and dangerous – fossils and images of former worlds became important elements woven into their poetry. Percy Bysshe Shelley is known to have owned Parkinson's works and many of the scientific references in *Prometheus Unbound*, such as 'the secrets of the Earth's deep heart', have been traced to them.[32] Lord Byron, in his poem *Don Juan*, is also referring to *Organic Remains of a Former World* when he laments:

> But let it go:—it will one day be found
> With other relics of 'a former world',
> When this world shall be former, underground,
> Thrown topsy-turvy, twisted, crisp'd, and curl'd,
> Baked, fried, or burnt, turn'd inside-out, or drown'd,
> Like all the worlds before, which have been hurl'd
> First out of, and then back again to chaos,
> The superstratum which will overlay us.

The fossil was key to understanding that Nature had endured for vast ages before man arrived to interpret it.[33] Thus Tennyson in *In Memoriam* is aware when he looks at fossils in the quarry face that he is seeing into the distant past. Seeking consolation in the fact that although Nature may be careless with individual lives, she is nevertheless 'careful of the type'; he then realises that even this is not true for species are also transitory:

> 'So careful of the type?' but no.
> From scarped cliff and quarried stone
> She cries, 'A thousand types are gone:
> I care for nothing, all shall go . . .'

Geology mattered because it was starting to explain the hitherto inexplicable and the static world of biblical history would never be the same again. As a greater understanding about the Earth and its early inhabitants emerged, James Parkinson and his *Organic Remains* provided a way for poet, collector and the general public to start making sense of it all.[34]

Underpinning Parkinson's detailed work on fossils, the exquisite drawings and his boundless enthusiasm for his science was his apparent belief in a good and benevolent God – even if the biblical account of the formation of the Earth required a rather liberal interpretation. But one of the reasons *Organic Remains* became so immensely popular is precisely because Parkinson did manage to integrate the geological facts with prevailing religious beliefs. He bridged the gap between the scientist and the collector, presenting the new geological evidence and rather frightening images of 'mutilated wrecks of former ages' within a religious context with which the general public felt comfortable.[35] It was a remarkable achievement.

12
The name of the father, and of the son

Most of you have children, and if you are not devoid of affection for them, pleasures beyond expression will be derived from instructing them.

James Parkinson, 1800
The Villager's Friend and Physician

My Friends,
I have devoted nearly thirty years of severe labour of body and of mind, to the promotion of your comfort and welfare; and, having been more eager to render myself capable, by study and observation, of performing the duties of my profession, than to wring from you your hard earned penny, I find myself, at the end of my labours, a poorer man than when I commenced them.[1]

So begins *The Villager's Friend and Physician*, the small book Parkinson penned back in 1800, purportedly written by a village apothecary coming to the end of his days. Without any hope of making provision for his old age, his only option, the apothecary claims, is to give up his profession and desist from those exertions 'which exhaust the energies of the mind, and rob it of its fortitude'. He resolves to devote himself 'to some rustic employment', convinced he will then suffer 'much less distressful fatigue . . . and mental anxieties, to which I have been so frequently a prey'. We seem to have caught our Mr Parkinson at a low moment. Perhaps he was having a bad attack of gout.

Those opening lines of the first edition (later editions omitted them) speak from Parkinson's heart and their frustrated, almost bitter, tone makes him sound disillusioned with life and weary of its drudgery. Although only 45 at the time, Parkinson had already devoted almost 30 years of his life to his 'laborious and harassing' profession and the strain was beginning to take its toll. Twelve years later, despite the success of *Organic Remains* and the fame it brought him, the need to bring in an income and the ever-increasing size of his practice meant he was still on the same treadmill. Fortunately, his son was now old enough to help out.

As the eldest surviving boy, John William Keys Parkinson

This illustration from the cover of *The Villager's Friend* shows the village apothecary, presumably James Parkinson, giving 'the alehouse sermon'. It is believed to have been drawn by Parkinson and is the only indication of what he might have looked like that has ever been found.

would have had little choice but to follow in his father's and grandfather's footsteps, eventually becoming heir to the practice. At the age of sixteen he had been a dressing pupil at the London Hospital, like his father before him, under the still practising and now famous surgeon *Sir* William Blizard.* The young John Parkinson had attended the Medical School established by Blizard at the London and, after training there for a year, had begun an apprenticeship with his father. In line with Parkinson's recommendations from his own experiences, John had gained anatomical and surgical skills in the hospital *before* starting his apprenticeship. Given the frustrations that James had suffered as an apprentice, shut behind the counter of the dispensary for seven years, he was determined John should not endure the same, and treated him more as an assistant than an apprentice.

By 1811, only two years after finishing his apprenticeship, John was sufficiently confident to write two medical papers, both of which were published in the *Transactions of the Medical and Chirurgical Society*.† As he was not yet a member of the Society, the papers were presented by his father. The first, 'A Case of Trismus, Successfully Treated', reported how John had treated a woman with lockjaw (tetanus).[2] Surviving this dreadful condition was rare since people usually died of exhaustion and dehydration, being unable to open their mouths to drink, so reporting a successful case of treatment was an important development. John first attended Mrs D., a woman of 50, along with Blizard, three weeks after she had had an accident and suffered a compound fracture to her leg. She was experiencing pain and stiffness in the back of her neck, a difficulty in swallowing, and an inability to open her mouth more than half an inch. As

* Blizard had been knighted in 1802.

† Chirurgical is an archaic term meaning related to surgery.

John and Blizard knew, these were the definitive symptoms of lockjaw, indicating she had suffered a long period of infection.

John first gave her small doses of tincture of laudanum every hour, which partly relieved the spasm of her jaws, enabling her to swallow and thus preventing her dehydrating, accompanied by five grains of calomel, as a laxative. But when a stool had not been produced by eleven o'clock the following morning, and she complained that her symptoms were not improving, a 'scruple'* each of jalap and scammony were added to the calomel – both powders were derived from plants and were powerful purgatives. When even this did not produce the desired effect, she was given a 'turpentine glyster with infusion of senna', administered as an enema.[3] This eventually resulted in the evacuation of 'a considerable quantity of dark coloured and highly fetid faeces'. Despite this, she still complained of a pain in her abdomen so the treatment was regularly repeated every three or four hours, seldom producing 'less than two copious stools in the day, but sometimes four or five'. For three weeks Mrs D. required constant medical care until, as her condition improved, her opium dosage was gradually reduced and five weeks later John Parkinson was able to report that she was returned to full health. Mrs D. may well have been the first person to ever recover from this condition, though quite how evacuating her bowels so regularly helped cure a tetanus infection is unclear.

Six months after recounting this case, John was ready with his second report, this time a case of acute appendicitis.[4] A boy of five had been observed for some time to be in declining health, but had made no particular complaint until two days before his death when he was suddenly seized with vomiting and appeared extremely weak. When John was called in he found the child's abdomen very swollen and painful on being pressed, and his

* A scruple was about 1.3 grams.

pulse hardly perceptible. 'Death, preceded by extreme restlessness and delirium, took place within twenty-four hours.' Unusually for the time, John performed an autopsy on the child and discovered that the cause of death had been a burst appendix.

After Parkinson again presented his son's paper, he was congratulated for this important contribution to medical science and John's account of this procedure has been claimed as the first written report of an acute appendicitis to appear in the English medical literature.[5] In particular, it was an important example of how an autopsy could reveal the cause of death – an indication that changes were beginning to occur in the profession.

Back in March 1805, in order to help raise funds for the ongoing Napoleonic Wars, the Government had increased the tax on materials used to make glass by 50 per cent, a measure which had brought in an additional £80,000 in duty.[6] By 1811, after a determined campaign by the glass manufacturers against this measure, the tax was finally removed from the raw materials and applied to the finished glass goods, which meant the tax was no longer paid by the manufacturers but by their customers. This threatened to greatly increase the price of medicine bottles that apothecaries used in large numbers, so on 3 July 1812, feeling that this tax was a particularly unfair burden on their profession, a number of apothecaries gathered at the Crown and Anchor Tavern on the Strand to express their concern. Father and son Parkinson were both present.

During the meeting, one of the speakers pointed out that while the tax on glass was undoubtedly an important issue, a more significant problem was the question of medical reform. His argument seemed to capture the mood of the moment and discussion turned to the regulation of their profession, which was overseen by the somewhat moribund Worshipful Society of Apothecaries. The apothecaries decided that they were not going to let things carry on as they were, so before disbanding they appointed a committee of twenty men, headed by George Man Burrows, which would investigate and report on apothecaries' grievances. Four months later Burrows delivered his report, which concluded that the problems were largely due to the lack of standardisation of training and qualifications. The immediate outcome was that the group disassociated themselves from the Society of Apothecaries and formed the rival Association of Apothecaries and Surgeon-Apothecaries of England and Wales, with George Burrows elected Chairman.[7]

The Association grew quickly – within a short time it had attracted over 1,000 members – and agitated for an Act to set up a body that would regulate general practice. Burrows worked hard towards this goal and later recalled that over the ensuing three years he attended 130 committee meetings and personally answered 1,500 letters. His efforts finally bore fruit in the Apothecaries' Act of 1815, which, by and large, regulated the licensing of apothecaries, although Burrows became deeply disillusioned at the nepotistic way in which the Act, especially its licensing procedure, was enforced by the Society of Apothecaries, the body disappointingly given responsibility for implementing the new procedures.

Many of the reformers' recommendations regarding medical education had been those expressed by Parkinson more than twelve years earlier in his book *The Hospital Pupil*, and it seems likely that Parkinson and Burrows collaborated on drawing up the report, for when Burrows retired from the presidency he was succeeded by Parkinson, who 'continued to discharge the duties of that very troublesome, but now less conspicuous office, with unwearied attention for two years'.[8] The committee on which Parkinson served particularly appreciated the 'zeal and ability' with which he had fulfilled his duties, as well as his 'urbanity and kind-heartedness which characterised his intercourse with the committee, and shed a lustre even on his acknowledged professional and general attainments'. In 1822, after Parkinson retired, he became one of the Association's two vice-presidents, but as its objectives had been achieved by then membership began to wane and the Association was disbanded in 1833.

Around the same time that the new Association of Apothecaries was formed, the Board of Trustees of the Parish of St Leonard's unanimously elected James Parkinson and his son to carry out the duties of 'Surgeon, Apothecary and Man-Midwife to the Poor of the Parish'. Their new role required the two men to visit sick paupers in their homes and to conduct a dispensary for those who were able to attend as outpatients. They would also be medical officers to the workhouse, which accommodated up to 800 inmates. It was an enormous amount of extra work to take on for the £156 a year they were paid between them. Even when, a year later, the salary was acknowledged to be 'not adequate to their services', and increased to 250 guineas, back-dated to 'Michaelmass last', it is difficult to see quite why they agreed to take on so much additional work. It may have been for altruistic reasons or, if the opening paragraphs of *The Villager's Friend* are to be believed, because they needed the extra income.

People would end up in the workhouse if they were too poor, too old or too sick to support themselves and in Shoreditch such unfortunates had been catered for in the parish workhouse from 1726. However, in 1774 the Shoreditch Vestry had been authorised by an Act of Parliament to levy a special Poor Rate for the purpose of building a new workhouse, which had opened in 1777. As soon as the Parkinsons accepted their appointment, they implemented major changes in the workhouse infirmary, which was now more than 30 years old, first separating the medical from the surgical cases, then rearranging the wards to include a maternity ward, a ward for chronic and incurable cases, and a ward for imbeciles.

A few months later, father and son had to deal with an outbreak of putrid fever, now called typhus, epidemics of which occurred repeatedly throughout Europe in the early 1800s. Epidemic typhus is spread by human lice – then considered more of an irritation than a health hazard – which become infected by feeding on the blood of patients with typhus fever. The disease spread rapidly through communities living in overcrowded and unhygienic tenements and workhouses, and has been one of the great scourges in human history. During Napoleon's retreat from Moscow in 1812, for example, more French soldiers died of typhus than were killed by the Russians. About ten days after being bitten, an infected person experiences headache, loss of appetite and a rapid rise in temperature, accompanied by fever, chills, extreme exhaustion and nausea. Four to six days after onset a characteristic rash appears over most of the body; the sick person is flushed, and the eyes are bleary. The temperature reaches a maximum by the end of the first week and is sustained until about the twelfth day. If the disease is untreated, circulation becomes sluggish and there may be spots of gangrene on the fingers, genitals, nose, and ears. Signs of pneumonia or kidney failure are common. Exhaustion is progressive, and delirium and coma follow; cardiac failure is usually the immediate cause of death.

As there was only one hospital in the whole of London specifically for the treatment of fever – the Fever Institution in Islington, which only had 70 beds – when epidemics occurred there were never enough facilities to cope. Parkinson expressed his astonishment that 'in an age distinguished for its flow of charity and kindly consideration of the poor' there was only one such institution 'for the reception of the pauper fever patients in London and its environs', and suggested that 'every parish [should] provide fever-wards for their own poor'.[9] Accordingly, the Parkinsons wrote to the Trustees of the Poor of the Parish of St Leonard's expressing their concern that in the past few months more than 100 people in the workhouse and surrounding tenements had died from putrid fever. They requested funds be made available to build a new fever ward so fever patients could be separated from the others, thus reducing the risk of infection.[10]

It took almost four years for the Trustees to make a decision, but eventually they approved the construction of a small building adjacent to the workhouse infirmary. It provided eighteen beds – six for men and twelve for women and children. This was the first separate fever ward to be set up in London in either a general hospital or a workhouse, and it is to the great credit of the Parkinsons that this innovation occurred, since senior physicians in many of the major hospitals still did not believe that typhus was contagious. Dr Thomas Young, physician to St George's Hospital, stated emphatically that he was not in favour of segregating fever cases: 'I should think it probable that the disease being more concentrated [in fever wards] would more easily be communicated to the nurses and physicians.'

In 1818 a second wave of the fever broke out, but this time only five people died in the St Leonard's workhouse, which clearly demonstrated the success of the new policy. As usual, James communicated his findings in a pamphlet entitled *Observations on the Necessity for Parochial Fever Wards, with remarks on the present extensive spread of fever*. Although no copy can be found today, it was reviewed at the time, providing us with an idea of its content.[11] The *London Medical and Physical Journal* particularly commended Parkinson's approach:

> the measure urged by Mr. Parkinson with so much laudable zeal, is sufficiently obvious. The numerous scenes of wretchedness that he has witnessed in consequence of the ravages of the prevalent fever . . . cannot fail to excite the attention of all those who may become acquainted with them, and lead to exertions for the prevention of similar calamities in future.[12]

While the *European Magazine* reflected that:

> Mr. Parkinson very judiciously recommends the revival of what have been called pest houses, for fever as well as small pox . . . The matter of this pamphlet is highly good; not false philanthropy.

Later that year, a committee was appointed by the House of Commons to enquire into the 'state of infectious fever in the Metropolis'. Not being physicians, neither of the Parkinsons were among those requested to give evidence to the committee; the majority who did attend still did not consider it important to separate fever patients from the others. Dr Alexander Marcet of Guy's Hospital and Dr John Yelloly of the London Hospital, both of whom were also members of the Geological Society and knew Parkinson well, were in complete agreement that it was *not* necessary to separate fever patients from the others, despite admitting that many patients, nurses, porters and visitors contracted the fever and died during epidemics. Yet again, it was decades before Parkinson's recommendations were routinely implemented.

One of the most common problems that Parkinson had to deal with on a daily basis was drunkenness, despite many of his medical works warning against its dangers. Because the quality of the water was so poor and could not be drunk, ale and porter were consumed in vast quantities in the 6,000 London alehouses, while the better off drank wine or port at home, often on a heroic scale. To be drunk was a natural and commonplace condition and Parkinson frequently had to attend the victims of drunkenness. One such incident ended with John Jones, 32 and a razor grinder, being indicted for murdering his wife. Both husband and wife were extremely drunk in an alehouse when they fell into an argument. She attempted to stop him leaving, tearing his clothes in the process, whereupon Jones punched her in the belly and she fell on to a pile of bricks. Witnesses said he struck her another violent blow while she lay there. The following day Parkinson was called to bleed her, but she died later on. In his autopsy report, Parkinson said he found her bladder was ruptured, her bowels inflamed and she was in a generally unhealthy state. He provided this evidence at the trial, but although the judge had already reduced the charge of murder to manslaughter, Jones was still found not guilty[13] – beating a wife when drunk was par for the course. Throughout his life Parkinson attended many such drunken incidents, along with several suicides and the occasional murder, but nothing quite prepared father and son for the shocking accident they were about to become involved with.

Parkinson's medical writings had made him well known across the City, and due to his conscientious and diligent approach to all health matters, his reputation and stature grew well beyond that of a normal apothecary-surgeon. As a consequence, Sir William Blizard called upon him one day to give his opinion in the extraordinary case of a Mr Tipple. Thirty-four-year-old Thomas Tipple had travelled in a single-horse chaise to visit friends in Forest Gate, then a village near Epping Forest. When he arrived at about nine o'clock in the evening, he proceeded to unharness the horse, but made a mistake in taking off the bridle before the horse had been released from the shaft of the chaise. Suddenly the horse bolted towards the stable door, with Mr Tipple standing in its path. The shaft pierced Tipple under the left arm, passed right through his body and out the other side, exiting under his right arm and pinning Tipple to the wooden wall opposite. The end of the shaft protruded five-and-a-half inches on the other side. Two passing servants heard Tipple's screams and within a couple of minutes they released him and withdrew the shaft from his body. In doing so, small pieces of flesh were left adhering to the wood along its length.

Tipple then walked unaided into the house, and with a steady step proceeded to the first floor where he was advised to stop. He remembered, however, that on a previous visit to this friend, he had been given a room on the second floor which was preferable as it looked east, was more airy and was free from the annoyance of the afternoon sun. He therefore continued up another flight of stairs unaided, before being helped into bed. Tipple remained in a sitting position until William Maiden, the local surgeon, arrived.[14] Maiden examined the wound and to his amazement felt air exiting from one of the holes in Tipple's chest, along with a considerable quantity of blood, indicating that there must be a hole in Tipple's lung. Tipple was now breathing with extreme difficulty, but when asked if he felt any pain he said, 'No; what I feel is a dreadful weight in my chest, as if I should be suffocated by the blood trickling on my lungs.'[15] Fearful that Tipple would suffocate from an internal haemorrhage, Maiden decided to bleed him until 'at least four pounds of blood were taken'.

Within two hours of the accident, Blizard arrived but advised Tipple to immediately settle his affairs, since he did not expect him to survive the night. To everyone's surprise, a week later Tipple was still alive. He still had 'distressing sensations in his chest' but there was less pain and he breathed more easily, so it was considered safe to remove his waistcoat and shirt, which had not been taken off previously while his condition was so critical. They had now become 'very unpleasant', due to much

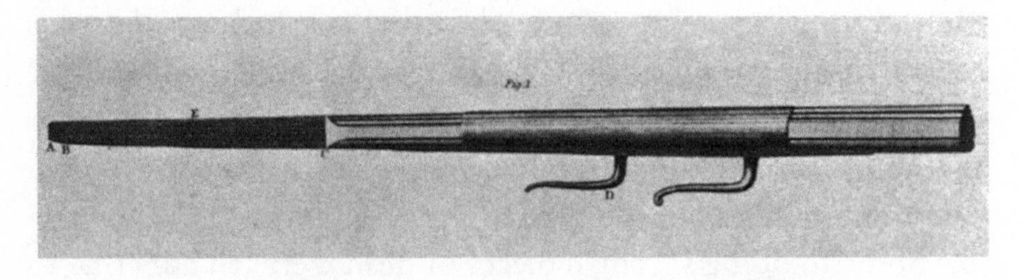

Sketch of the chaise shaft that pierced Mr Tipple.

dried blood adhering to them. Blizard still considered that Tipple would not survive but nevertheless decided to 'continue the struggle', feeling that, whatever the outcome, 'we shall have the consolatory reflection of having done our duty'.[16]

Four weeks later Tipple was able to walk and well enough to return to his home in Hoxton. Blizard generously attributed the patient's remarkable recovery to Maiden's treatment which consisted of copious bleeding – eleven-and-a-half-pounds of blood were removed in total – a low diet and purgatives, and especially Maiden's 'judicious forbearance of the use of a finger or a probe for the purpose of ascertaining the course of the shaft'. Many lives, he noted, had been sacrificed to the gratification of curiosity by probing the direction and extent of wounds in the body – often, no doubt, with dirty fingers or instruments.[17]

The case was so unprecedented that Blizard's report of Tipple's recovery, sent to the Royal College of Surgeons, was not at first believed, several surgeons considering that the injuries could not have been as severe as Blizard made out. In order to remove any doubt, Maiden had the testimonies of Tipple and witnesses sworn before the local magistrates who had investigated the accident. Even so, there were still some who questioned Blizard's account. Georgian society was obsessed with respectability and reputation, so someone as distinguished as Sir William, now 70, would have felt the slight to his character very keenly. Consequently, three months after the accident, he called a meeting of 'eighteen of the principal medical men in London' to examine the wounds before they were completely healed. This group of eminent surgeons included both father and son Parkinson, as well as Astley Cooper, Henry Cline and John Abernethy. At the end of the examination, all eighteen gentlemen were quite satisfied that the shaft had gone right through the cavity of Tipple's chest, from left to right.

When Tipple died eleven years later, John Parkinson had been attending him for five years, but it was Mrs Tipple who suggested that an autopsy be performed. The examination was carried out under Blizard's supervision by William Clift, the curator of John Hunter's Museum at the Royal College of Surgeons. They were assisted by John Parkinson 'in the presence of several other professional Men', including James Parkinson. What they discovered was that Tipple still had a number of broken ribs which protruded into his chest cavity, and two large holes either side of his sternum marked the entrance and exit points of the chaise shaft. In addition, his chest muscles were badly damaged and showed substantial scarring, and his lungs were an unusual colour and of a particularly dense texture, which the medical men deduced must have caused the restriction in his breathing.

While the post-mortem was conducted largely on account of this case being so unusual, it seems another reason was to

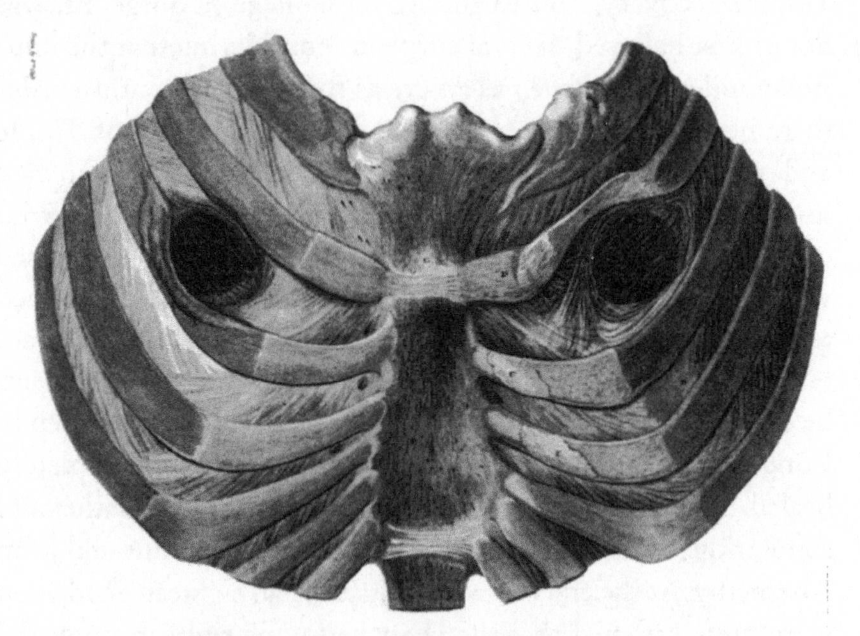

William Clift's drawing of Mr Tipple's rib cage, showing the entrance (right side when looking at the drawing) and exit points of the chaise shaft.

ascertain once and for all that the shaft had penetrated right through Tipple's body. Publication of the post-mortem report, and William Clift's superb drawings, at last fully exonerated Blizard from any possibility that he had exaggerated the man's wounds. The shaft of the chaise and Mr Tipple's ribcage were subsequently presented by Maiden to the Royal College of Surgeons 'as a trophy of modern surgery'. The items were then deposited in the Hunterian Museum 'for the inspection of the curious', where they remained as notable exhibits for more than a century until they were destroyed by German bombs in 1941 – along with a number of Parkinson's fossils.[18]

It must have been flattering for Parkinson to be considered, shoulder to shoulder with his son, among the most important medical practitioners in London. He had achieved a great many things in his life, and had been gratifyingly recognised for them. For some, that would have meant it was time to slow down. Not Parkinson. He continued to work and continued to write in his spare time. Indeed, he still had a great deal to do; he still had his most famous paper to write.

13
The shaking palsy

The name Parkinson's disease has one major advantage over the term 'paralysis agitans' – it does not frighten the patient!

R. Williamson, 1925
James Parkinson and his essay on paralysis agitans

IN HIS NOVEL *Humphry Clinker*, the Scottish surgeon and novelist Tobias Smollett described London in 1771 as an 'overgrown monster', when referring to the rapid increase in its population and the effect that had on daily life:

> . . . they are seen every where, rambling, riding, rolling, rushing, justling, mixing, bouncing, cracking, and crashing in one vile ferment of stupidity and corruption—All is tumult and hurry; one would imagine they were impelled by some disorder of the brain, that will not suffer them to be at rest.

Forty years later, even the once-peaceful village of Hoxton had been swallowed up by this leviathan, its quiet market gardens dug up for housing to accommodate ever-increasing numbers; its narrow streets a thoroughfare for those on their way from the country to the city for work, trade or business. The place was dirty, noisy and overcrowded – so the middle classes fled the disease and immorality they associated with the poor, and charitable institutions took over the old houses they left behind, turning them into asylums, schools, almshouses, hospitals, madhouses or workhouses. Light industry also began to overwhelm Hoxton, as exemplified by the greatly enlarged brickfields, where clay had been quarried since at least Tudor times, and which dominated the landscape at the northern end of Kingsland Road.

The streets around Hoxton Square, with which Parkinson was so familiar, were busy from morning till night, particularly around Hoxton Market, to the rear of his house. Over the years, going about his daily business amid the hustle and bustle, he would often see the same faces again and again, instinctively taking note as their diseases ebbed and flowed. Many of those he observed were elderly and infirm as the surrounding area had an unusually high number of almshouses and workhouses. Three of these individuals particularly attracted his attention because they all showed symptoms similar to those he had encountered in a patient a number of years previously – they were all men whose extremities shook alarmingly; they also tended to walk or run on the balls of their feet and would have fallen forward had they not been supported either by a stick or by an attendant. It was a curious condition and one Parkinson was very interested to know more about, so he stopped each of them on the street and enquired after their health. Each man was considerably stooped, his speech was slurred and difficult to understand, and they all told Parkinson that their condition had come on gradually.

Over time two other cases with similar symptoms came to Parkinson's attention and after some six years of making notes on the condition and deterioration of each man, he decided to share his observations with the world. Although somewhat tentative about publishing, since he was unable to offer an explanation or cure for the disease, he nevertheless considered he would be fully rewarded if he achieved nothing else but 'excited the attention of those who may point out the means of relieving a tedious and most distressing malady'.

James Parkinson's treatise, *An Essay on the Shaking Palsy*, published in 1817 when Parkinson was 62, has deservedly become a medical classic.[1] Original copies of the work are now rare, although facsimiles have been reproduced from time to time and scanned versions of the original can be found online.[2] Its significance lies in the fact that Parkinson was the first to identify and describe the symptoms that defined the shaking palsy, known to us today as Parkinson's disease.

The *Essay* opens with Parkinson's famous definition of the shaking palsy, in which he captures the very essence of the disease:

> SHAKING PALSY. *(Paralysis Agitans.)*
> Involuntary tremulous motion, with lessened muscular power, in parts [limbs] not in action and even when supported; with a propensity to bend the trunk forward, and to pass from a walking to a running pace: the senses and intellects being uninjured.

As with so much of Parkinson's work, it was not only his acute observational powers and attention to detail that enabled him to provide such an insightful commentary on the shaking palsy; he also researched the medical literature extensively before committing his ideas to paper. So as well as citing Galen, he refers to the works of the Dutch physician Sylvius de la Boë (who published in 1680), the German physician Johann Juncker (1723), the Dutch-Austrian physician Gerard van Swieten (1745), Hieronymus David Gaubius, a German physician and chemist (1758), and the French physician and botanist François Boissier de Sauvages (1768), quoting from each in the language in which they wrote. All these writers had identified various symptoms of the shaking palsy, such as the propensity to run when the patient intended only to walk, and some even distinguished different kinds of tremor, but none of them recognised that all these disparate symptoms resulted from a single disease, as Parkinson was to do.

Sauvages, for example, had described what is known as the resting tremor *and* the unusual means of walking, but he had not recognised that the two symptoms were characteristic of the same disease. Parkinson on the other hand was fully aware that he was describing a disease which resulted from an assemblage of symptoms, 'some of which do not appear to have yet engaged the general notice of the profession'. As he explained, the term 'shaking palsy' had been used to loosely describe various conditions: 'By some it has been used to designate ordinary cases of Palsy, in which some slight tremblings have occurred; whilst by others it has been applied to certain anomalous affections, not belonging to Palsy', so in Chapter 2 he gives examples of where other doctors had diagnosed the shaking palsy incorrectly. There was the woman who constantly swung her arm 'like the swing of a pendulum'; the young boy with worms whose legs had become useless; the tremblings caused by the drinking of spirituous liquors and those which proceed from the 'immoderate employment of tea and coffee'; as well as the shaking that came on with old age and infirmity. All these different tremors were at one time incorrectly ascribed to the shaking palsy, but it was Parkinson who first recognised the crucial difference between these and the *real* shaking palsy:

> If the trembling limb be supported, and none of its muscles be called into action, the trembling will cease. In the real Shaking Palsy the reverse of this takes place, the agitation continues in full force whilst the limb is at rest and unemployed . . .

In other words, if the patient placed a trembling arm on the table and it stopped trembling, then they did not have the real shaking palsy – a mistake even the great John Hunter made when describing the condition. Only if a trembling arm resting on the table continued to shake could the tremor be ascribed to the shaking palsy. This resting or 'Parkinsonian' tremor is now one of the three diagnostic symptoms of Parkinson's disease, occurring in 70–80 per cent of cases, and it is usually the first sign of the condition.* Once he had recognised that this resting tremor (Parkinson called it 'palpitation of the limbs') was a diagnostic symptom, he then pointed out why it was necessary to separate it from other forms of tremor: '[because] the distinction may assist in leading to a knowledge of the seat of the disease'.

Having observed the first of his patients with the condition over a period of about six years, Parkinson noted the different stages the man went through and was able to establish a history of how the disease progressed. He recognised that it was 'of long duration' and that the various stages manifested different symptoms, so when he encountered the other men displaying some of those symptoms, he realised they were all suffering from the same condition, but were at different phases in its progression. This allowed him to describe how the shaking palsy inexorably moved from one stage to another:

> So slight and nearly imperceptible are the first inroads of this malady, and so extremely slow is its progress, that it rarely happens that the patient can form any recollection of the precise period of its commencement. The first symptoms perceived are a slight sense of weakness, with a proneness to trembling in some particular part; sometimes in the head, but most commonly in one of the hands and arms.

Indeed, in the case of Mike Robins, whom we met in the Prologue, the condition started exactly as Parkinson describes,

* The other two definitive symptoms of Parkinson's disease are muscular rigidity and slowness of movement.

with a barely perceptible twitch in Mike's right shoulder which within a few months had become an uncontrollable tremor down his right arm.

Parkinson continues:

> After a few more months the patient is found to be less strict than usual in preserving an upright posture; this being most observable whilst walking, but sometimes whilst sitting or standing. Sometime after the appearance of this symptom, and during its slow increase, one of the legs is discovered slightly to tremble, and is also found to suffer fatigue sooner than the leg on the other side. And in a few months this limb becomes agitated by similar tremblings, and suffers a similar loss of power.

As the disease proceeds, familiar tasks are accomplished with greater and greater difficulty,

> ... the hand failing to answer with exactness the dictates of the will. Walking becomes a task which cannot be performed without considerable attention. The legs are not raised to that height, or with that promptitude which the will directs, so that the utmost care is necessary to prevent frequent falls.
>
> ... at meals the fork, not being duly directed, frequently fails to raise the morsel from the plate which, when seized, is with much difficulty conveyed to the mouth.

At this point the patient rarely gets any respite from the shaking:

> Commencing, for instance in one arm, the wearisome agitation is borne until beyond sufferance, when by suddenly changing the posture it is for a time stopped in that limb, [only] to commence, generally, in less than a minute in one of the legs, or in the arm of the other side.

Gradually things get worse:

> In some cases . . . the patient can no longer exercise himself by walking in his usual manner, but is thrown on the toes and fore part of the feet; being, at the same time, irresistibly impelled to take much quicker and shorter steps, and thereby to adopt unwillingly a running pace . . . since otherwise the patient, on proceeding only a very few paces, would inevitably fall.
>
> In this stage, the sleep becomes much disturbed. The tremulous motion of the limbs occur during sleep, and augment until they awaken the patient, and frequently with much agitation and alarm.

This comment regarding sleep is the only inaccurate part of Parkinson's description. While patients may have problems falling asleep because they are anxious or depressed, or staying asleep because they cannot turn over or change position in bed due to their rigidity, they do not shake during deep sleep. It is unlikely that James ever had the opportunity to observe anyone asleep who had the shaking palsy, and because patients themselves thought they shook while asleep – since on waking they immediately began to shake – they reported to Parkinson that they had been shaking all night.

Eventually, Parkinson's description continues, the patient is unable to feed himself, defecate, walk, speak or chew, such that saliva 'is continually draining from the mouth, mixed with the particles of food, which he is no longer able to clear from the inside of the mouth'. Ultimately, 'constant sleepiness, with slight delirium, and other marks of extreme exhaustion, announce the wished-for release'. It was a terrible disease with appalling symptoms that nothing could relieve except death.

But being able to identify the symptoms was only part of the problem – far more difficult was finding the cause of the condition. In an attempt to understand this, Parkinson enquired of the sufferers he encountered as to when and how their symptoms had started and whether they considered that anything in particular had triggered the onset of the disease. His first patient had been a gardener all his life and thought his problems had arisen after he was engaged for several days in employment that required considerable exertion of his left arm, which was when the symptoms had first appeared; otherwise he had led 'a life of remarkable temperance and sobriety'. Over a number of years this man had slowly deteriorated and it was his physician's lack of understanding of the disease that had first aroused Parkinson's interest in it.

The second man had been an attendant in a magistrate's office and recalled the disease coming on slowly, attributing it to 'considerable irregularities in his mode of living, and particularly of indulgence in spirituous liquors'. The third, a sailor 'of a remarkable athletic frame', thought his affliction had started after a spell of several months lying on the bare, damp earth of a Spanish prison. Case number four moved to a separate part of the country before Parkinson could ascertain many details of his disease, and number five was only seen at a distance running on the balls of his feet, his attendant running backwards in front of him, ready to catch him should he fall. The sixth and final patient observed by Parkinson was a particularly interesting case. He had led a remarkably temperate life and could not point to any situation or circumstance which he considered had triggered the disease. It had begun about twelve years previously and was now so bad that one limb or other shook almost permanently. The shaking would start gradually but would eventually become so violent the whole room would shake. With a great deal of effort he was able to change his position and the shaking would cease for a few minutes, but soon afterwards it began again in another limb. He told Parkinson that a year previously he had suffered a stroke during the night and was temporarily paralysed down his right side. The whole time he remained paralysed he did not shake, but within two weeks of regaining the use of his arm and leg, the shaking returned.

From this questioning of his patients it was evident to Parkinson that no specific lifestyle factors had caused the onset of the disease, making the condition even more perplexing as to its cause. He was nevertheless still convinced that 'The preceding cases appear to belong to the same species'.[3] But if their mode of living were not to blame, what was? In Chapter 4 of his *Essay* Parkinson considered this interesting question, but before doing so he prefaced his suggestions with many caveats:

> Before making the attempt to point out the nature and cause of this disease, it is necessary to plead, that it is made under very unfavourable circumstances . . . Not having had the advantage, in a single case, of that light which anatomical examination yields, opinions and not facts can only be offered. Conjecture founded on analogy, and an attentive consideration of the peculiar symptoms of the disease, have been the only guides that could be obtained for this research, the results of which is, as it ought to be, offered with hesitation.

Parkinson believed that the source of the problem could not be the brain because of the 'absence of any injury to the senses and the intellect'. In fact, about 30 per cent of cases do result in some form of dementia, but his patients probably died before any dementia had time to manifest itself. So because all the symptoms seemed to be of an involuntary nature, and the whole body eventually became afflicted with the disease, he reasoned that the cause must lie in the spinal canal from which the spinal nerves emerged, rather than in the nerves themselves. Perhaps the patients had suffered an injury to the neck at some point in their past? None could remember such an event. Or some kind of rheumatic attack that had occurred so long ago, they had forgotten about it? He then discussed a number of cases believed to have been caused by some injury to the neck or spine and considered the possibility that some of these might have developed into the shaking palsy. But having examined the facts he was forced to doubt 'the probability of its being the direct effect of any sudden injury'. He therefore concluded that

> taking all circumstances into due consideration, particularly the very gradual manner in which the disease commences . . . as well as an inability to ascribe its origin to any more obvious cause, we are led to seek for [a cause] in some slow morbid change in the structure of the medulla [oblongata] . . . occasioned by simple inflammation or rheumatic or scrophulous affection.

Such damage, he thought, might gradually affect the upper part of the spinal cord (*medulla spinalis*), which then spread into the lower part of the brainstem (*medulla oblongata*) as the disease progressed. In fact, injury to the medulla oblongata may result in a number of sensory related problems, such as paralysis, difficulty swallowing and lack of movement control, so his reasoning for its time was quite astute.

The final chapter, 'Considerations respecting the means of cure', opens rather despondently: 'The enquiries made in the preceding pages yield, it is much to be regretted, but little more than evidence of inference: nothing direct and satisfactory has been obtained.' Nevertheless, he considers it should not be assumed that a cure cannot be found. 'On the contrary,' he declares more optimistically, 'there appears to be sufficient reason for hoping that some remedial process may ere long be discovered, by which, at least, the progress of the disease may be stopped.' He never dreamt it would take almost 200 years.

As one case he had heard about appeared to have responded to the treatment that had been administered, he considered that even if there was only the smallest chance of a cure, that treatment should be given a trial. He therefore advised the taking of blood from the upper part of the neck, since that was where he considered the problem lay, after which blistering should be applied to the same place until a 'purulent discharge' (pus) was obtained. If, on carrying out this treatment, an insufficient discharge was achieved, he further recommended that blisters 'of at least an inch and a half in length might be established' on either side of the upper spine. These blisters were best made with 'caustic' (probably sodium hydroxide) and kept open with any 'proper substance'; cork was ideal, being light and soft but firm and elastic, and capable of being readily fashioned into any convenient shape. It is perhaps not surprising that on proposing this treatment to one of the sufferers he met in the street, his offer was declined.

In the rationale for his diagnosis and therapy, Parkinson was acutely aware 'that mere conjecture takes the place of experiment' and that anatomical examination was the only way forward, but he did not have facilities in his laboratory to carry out autopsies. Although autopsies had become more common after Giovanni Morgagni's *The Seats and Causes of Diseases Investigated by Anatomy* had been translated into English in 1769, they were by no means routine and there were no formal procedures with which to describe any clinical features in detail. It was not until 1874 that a German doctor, Rudolf Virchow, introduced a standardised technique for performing autopsies, during which the whole body was examined in detail, often revealing unsuspected problems.

Parkinson was an acute observer, which enabled him to discover associations not previously recognised by others, but he also recognised that in order to confirm such relationships, it was necessary to carry out experiments to verify his assumptions. His appeal for more autopsies to be performed on people who had died from the shaking palsy is repeated in the closing pages where he states that an important reason for writing his *Essay* was to draw the disease to the attention of 'those who humanely employ anatomical examination', since it was only through autopsy that the real nature of the disease could be ascertained, 'and appropriate modes of relief, or even of cure, pointed out'. As there was still much public opposition to cutting up dead bodies, he concludes with a justification of the practice, explaining how medical science was already deeply indebted to those who performed such research:

> Little is the public aware of the obligations it owes to those who, led by professional ardour, and the dictates of duty, have devoted themselves to these pursuits, under circumstances most unpleasant and forbidding.

Today one could say the same about researchers who have to perform medical experiments on animals, against much public resistance. Various drugs, as well as operations like deep brain stimulation, now provide considerable relief to sufferers of Parkinson's disease. But none of this could have been achieved without trialling them on animals first, which no researcher likes to do.

The *Essay* first appeared in early June 1817.[4] It was favourably reviewed in the medical press, although it was certainly not recognised at the time as the seminal work it turned out to be. *The Medico-Chirurgical Journal and Review* reassured Parkinson that he should not fear putting forward his ideas, even though they were just conjecture: 'The name of the author would be a sufficient passport to publicity, and security from aspersion, for a much less reputable performance.' *The Monthly Gazette of Health* concurred with this assessment of his reputation: 'The surgeon Parkinson, a practitioner of considerable experience and scientific attainment has lately published a popular treatise on this disease', adding, 'but although at present uninformed as to the precise nature of the disease, still it ought not to be considered as one against which there is no countervailing remedy.'[5]

Parkinson wrote about the symptoms he observed with such accuracy that even though he could not identify its cause, he enabled others who came after him to use his observations on which to build their understanding. Despite this, little progress was made for many years and there seems to have been scant reference to the *Essay* within the medical literature until some six years after his death (thirteen years after the *Essay* was written), when Dr John Elliotson mentioned it in a lecture delivered at St Thomas's Hospital in 1830:

> The best account of this disease which I have seen is one given by a general practitioner, now deceased, of the name of Parkinson, a highly respectable man, who wrote an essay upon the subject in 1817, from which I have derived nearly all I know upon the complaint . . . It does not appear as if this disease of which I am at present speaking, was well characterised or distinguished before Mr. Parkinson wrote on the subject.[6]

More than ten years after this, the situation had changed little. Sir Thomas Watson, lecturing on *paralysis agitans*, as the shaking palsy was also known, at King's College London in 1841, told his students:

> Allusions to this form of disease are to be found in many of the older systematic writers on physic; but it never was much attended to in this country until Mr. Parkinson published an essay upon it in the year 1817; and a very interesting little pamphlet it is.[7]

Watson, however, seemed nonplussed as to how to deal with the 'state of decay', as he described the disease, other than to 'regulate the bowels, to procure sleep, to nourish and uphold the patient without unduly stimulating him; and this is all that I can tell you of the shaking palsy'.

It was not until the 1860s that Parkinson's *Essay* became more widely cited. Daniel Maclachlan in his book *A Practical Treatise on the Diseases and Infirmities of Advanced Life*, written in 1863, refers to 'Mr. Parkinson whose interesting essay must ever be referred to, as giving a faithful account of the symptoms of the disease from beginning to the end, and is still the best work on the subject'.[8] Two years later, William Sanders, in a paper on an unusual case of nervous disease, which he called 'pseudo-paralysis agitans', also referred to Parkinson's work and implied others were now following Parkinson's classification of the symptoms. When going on to discuss a more appropriate name for the shaking palsy or *paralysis agitans*, Sanders refers to it as 'Parkinson's disease', but in doing so he means the disease as described by Parkinson, and is not suggesting it should be called that. Along with several alternative names, he does propose *paralysis agitans Parkinsonii*, but this rather cumbersome mouthful did not catch on.[9]

It was more than 50 years after publication of Parkinson's *Essay* before anyone seriously turned their attention to the disease, as Thomas Buzzard pointed out in 1882:

> The disease 'shaking palsy', or '*paralysis agitans*' . . . was first regularly described by our countryman Parkinson in 1817. Parkinson was a member of the Royal College of Surgeons, and his *Essay on the Shaking Palsy* presents so graphic and admirable a description of the disease that comparatively little has been left for subsequent observers to add to his account. In our time Charcot has also made the disease the subject of clinical investigation.[10]

Jean-Martin Charcot was a French neurologist who worked and taught at the famous Salpêtrière Hospital in Paris.[11] In 1882 he opened Europe's first neurology clinic at the Salpêtrière and wrote a series of articles that gained him a worldwide reputation in neurology. Charcot's detailed descriptions of Parkinson's disease are among his major contributions; in particular he distinguished slowness of movement (bradykinesia) as a feature of the disease,[12] which Parkinson had not identified as being symptomatic.

After a long and frustrating search, and with the help of an English librarian, Charcot had eventually managed to acquire a copy of Parkinson's *Essay*, which was yet to be translated into French. He wanted a copy because he considered it such a 'descriptive and vivid definition that is correct for many cases, most in fact, and will always have the advantage over others of having been the first'. Charcot also recognised what was perhaps Parkinson's greatest quality, his ability to observe:

> Let someone say of a doctor that he really knows his physiology or anatomy . . . these are not real compliments; but, if you say he is an observer, a man who knows how to see, this is perhaps the greatest compliment one can make.[13]

However, he felt that Parkinson had overlooked one particular characteristic of the disease: 'we allude to the rigidity to be found, at a certain stage of the disease in the muscles of the extremities of the body and for the most part in those of the neck also'.[14] While it is true that Parkinson does not use the word 'rigid', he does describe how the upper body is thrown so far forward that the patient is forced to run when he only wants to walk, which implies a rigid posture. Indeed, Charcot acknowledged that Parkinson described how the head is greatly bent forward, due to the muscles in the neck becoming rigid.

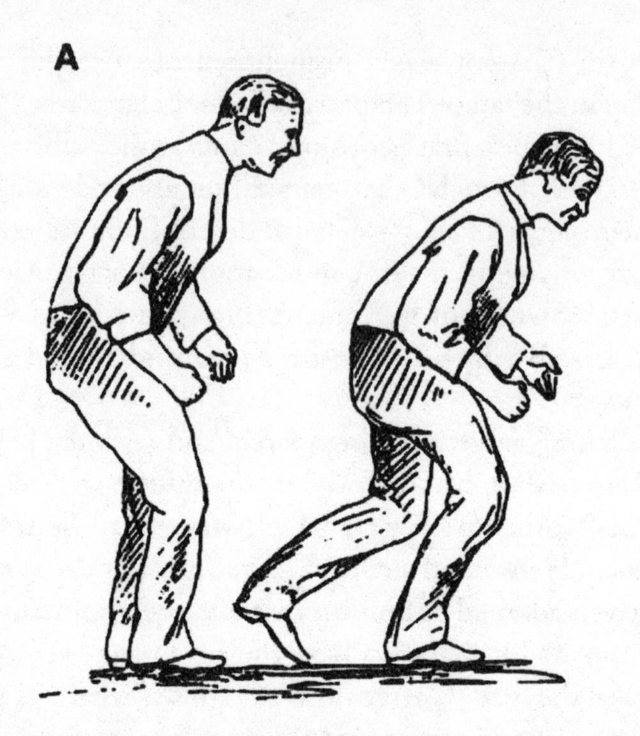

Drawing from Charcot's lesson given on 12 June 1888, which demonstrates the typical flexed posture of Parkinson's disease.

In 1872, when giving a lecture on *paralysis agitans*, Charcot explained to his students that 'The first regular description of it only dates from 1817; it is due to Dr Parkinson who published a little work entitled *Essay on the Shaking Palsy*'.[15] He also remarked how existing names for the disease were inappropriate since patients were not markedly weak or paralysed (*paralysis agitans*), but neither did they always have tremor, so the term shaking palsy was also unsuitable; it was then he suggested using the term *maladie de Parkinson* to describe the condition. 'Parkinson's disease', he considered, was a much more appropriate name.

Charcot refined and expanded Parkinson's early descriptions of the disease, and by the end of the nineteenth century the term 'Parkinson's disease' had become established in the medical literature. However, the cause and location of the affliction was still thought to be much as Parkinson had originally described it – some kind of damage or disease in the spinal cord or lower brainstem. It was to be another 50 years or so, in the mid-1950s, before it was recognised that cell damage in the substantia nigra (a structure in the mid-brain) was the cause.[16] Nerve cells (neurons) in the substantia nigra send out fibres to tissue located in both sides of the brain. These neurons normally produce the neurotransmitter dopamine, which sends signals to the basal ganglia, a mass of nerve fibres that helps to initiate and control movement. When the cells of the substantia nigra deteriorate, as in Parkinson's disease, there is a corresponding decrease in the amount of dopamine produced. The decreased levels of dopamine cause the neurons to fire uncontrollably, preventing the patient from being able to direct motor function.

Parkinson would have been very interested to know that. As he predicted, 'anatomical investigation' did finally reveal the true nature and location of the disease, and the 'benevolent labours' of subsequent research has meant that many treatments are now available to remediate the symptoms of this miserable condition. But even today there is no cure and the cause of cell deterioration in the substantia nigra is largely unknown. Let's hope it doesn't take another 200 years to find out.

14
Reforms and rewards

Reward follows, sooner or later, every just deed.
Sir William Blizard, 1823
Presentation of Gold Medal to James Parkinson

By the time the Napoleonic Wars ended in 1815, England was experiencing great social, economic and political upheaval, many of the problems caused by the speed of change in working conditions resulting from the Industrial Revolution. The Luddites, in particular, were trying to save their livelihoods – new-fangled machinery greatly reduced the number of people needing to be employed, threw thousands out of work, and left many without a legitimate means of earning a living. When a bad harvest exacerbated the situation, distress reached its climax and the Luddites started smashing industrial machines developed for use in the textile industries of the north. These incidents were frequently followed by execution of the culprits, enraging the workforce even further, and once again reform societies flourished demanding votes for all men.

On 15 November 1816, the radical speaker Henry Hunt, known as 'Orator' Hunt for his rabble-rousing speeches, was invited to address a mass meeting at the Spa Fields in Islington, London. The intention was to petition the Prince Regent, the future King George IV, to reform Parliament. In the twenty years since the demise of the London Corresponding Society, the agenda had not changed: the petition demanded universal suffrage, annual elections and secret ballots. The meeting was attended by about 10,000 people and Hunt spoke from the window of a public house wearing a white top hat, a symbol of radicalism and the 'purity of his cause'. He complained about the evils of high prices and over-taxation, and reiterated the need for parliamentary reform. The meeting dispersed peacefully and Hunt subsequently made two attempts to hand in the petition to the Prince Regent; on both occasions it was rejected.

On 2 December Hunt held a second demonstration at the Spa Fields to protest at the treatment he had received from the Prince. Although the meeting started peacefully enough, this time the crowd was swollen by people returning from a public hanging at Newgate Prison, many of whom were very drunk. A breakaway group of protesters began 'a scene of outrage and tumult',[1] which degenerated into what has become known as the Spa Fields Riot. One group broke into a gunsmith's shop, stole weapons and shot a man in the shop. Another group attempted to attack the Tower of London. The authorities, however, alerted by their spies, had plenty of constables, troops and horses on the alert and the ringleaders were eventually captured. One man was later hanged for committing a robbery during the riot and three others were charged with High Treason for plotting to kill the King and take control of the Government by seizing the Tower of London and the Bank of England. But when it transpired that most of the evidence against them was based on spy and informer testimony they were acquitted.

A month later, the much-hated Prince Regent, the Prince of 'Whales',* had the windows of his carriage smashed by a mob

* The Prince Regent was also the Prince of Wales, and he was very fat.

after the State Opening of Parliament. Believing that revolution was imminent, Parliament suspended *Habeas Corpus*, banned meetings of more than 50 people, and the Lord Lieutenants were again ordered to apprehend all 'printers, writers and demagogues responsible for seditious and blasphemous material'.[2]

These events must have reminded Parkinson of everything he had been through during the 1790s. It was now more than 25 years since he had first joined the London Corresponding Society, but his zeal for the cause of parliamentary reform remained undiminished.

Parkinson was still a Trustee of the Vestry for the Liberty of Hoxton, and although the Vestry's paramount concerns were the affairs of its parishioners on local matters, on occasions it organised meetings where the townspeople could protest about grievances affecting the nation. Since few parishioners had a parliamentary vote, the views of the people could only be vocalised by petitioning Parliament, which the Vestry facilitated.

At one such meeting of Hoxton parishioners on 13 February 1817 it was agreed that certain gentlemen form a committee 'authorised to carry into effect the above resolutions in such a manner as may appear to them most expedient'. Messrs Sharpe, Flower, Jacobson, Lawrence and Parkinson were duly elected to form the committee, which agreed to find a way of 'obtaining a redress of the numerous grievances under which the country groans and a full real and true representation in Parliament of such portions of the English people as are either misrepresented or not represented'.[3] The 'above resolutions' are peppered with phrases such as 'That we most solemnly deprecate the practice of rejecting the petitions of the people merely because the language is not suited to the Ears of those members whose Corrupt Practice we hold in the greatest abhorrence' and 'our unhappy Country has a long time born the Yoke of a Tyrannic Oligarchy'. The hand of James Parkinson can be read in every line.

The committee resolved that the initial course of action should be to send in a petition stating their concerns but, like the other petitions that were daily delivered to the Prince, it was to have no effect.

It was not until eight years after Parkinson's death that the Reform Bill finally arrived in 1832 and, when it came, many were disappointed. Property qualifications still meant that only one in seven adult males was eligible to vote. The growth and influence of the Chartist movement from 1838 onwards continued to pressure the Government for parliamentary reform, although it was another 30 years before the 1867 Reform Act gave the vote to every male adult householder living in a borough constituency, and to male lodgers paying at least £10 a year for unfurnished rooms. This brought the number entitled to vote to about 1.5 million. Seventeen years later, in 1884, men living in the country were given the same franchise as the boroughs, bringing the total number of voters to around 6 million. In 1918, at the end of the First World War in which almost a million young British and colonial men lost their lives, the 'Representation of the People Act' extended the vote to all men over the age of 21 and to women over 30. Finally, in 1928, universal suffrage was extended to women over the age of 21. Parliamentary reform was a slow process and despite all his efforts and those of his friends, Parkinson did not live to see it happen.

In 1818 the Reverend Dr William Buckland was appointed Reader in Geology, a post that had been created for him, at Oxford University. Buckland was not only a theologian, eventually becoming Dean of Westminster, but he was also one of the greatest geologists of his day. In his inaugural lecture, given in 1819, Buckland explained why the new science of geology was now included in the university's curriculum, alongside the established, highly religious and largely classical syllabus conventionally taught there. He also attempted to reconcile the mounting geological evidence with the biblical account of Creation. The lecture was published in 1820 under the title of *Vindiciae Geologiae; or the Connection of Geology with Religion Explained*, and Buckland kindly sent Parkinson a copy. Parkinson, normally highly courteous, took some time to respond because, as he eventually explained, he didn't feel '*quite* satisfied' with the lecture.

When he did reply, in January 1821, Parkinson pointed out that although Buckland had demonstrated that the planet was very old, the human race quite young, that fossils were often quite different from their living counterparts, and that all these facts were 'not compatible with the account of the Creation as given by Moses'; Buckland had omitted one crucial fact: he had not mentioned that no remains of the human race had ever been found in sediments thought to result from the Deluge.[4] Irritated that Buckland had left out this important point, and suspicious that he had done so deliberately, Parkinson was forthright in pointing out the consequences of such an omission, accusing Buckland of misdirecting 'the exertions of science', and appealing to his better judgement: 'When I look at the considerable degree of accordance which your labours have already made manifest I am impelled to wish that you would extend them to this point . . . that man was not created until after the deluge'.

For Parkinson the geological evidence was now beyond doubt and it was time to find an alternative explanation for the biblical account of Creation. This, he felt, did not diminish the fact that 'God is the creator of the world, and of everything contained in it'; it was simply that Moses' version of Creation as recorded in the Bible was inadequate, as he explained to Buckland:

> ... the account of the deluge may be attributed to his [Moses] ... uninformed judgement, embracing the adopted tradition of that day.

Parkinson's more sophisticated interpretation of Creation as revealed in this letter, compared with the views expressed in his books – in which he had refrained from making controversial claims that challenged the scriptures – probably came about because he felt able to express his true beliefs in private, whereas when writing for the general public he felt he needed to be more circumspect and explain things to his audience in terms he thought they wanted to hear. Unfortunately, we do not have Buckland's reply to this sharp reprimand, but he must have been somewhat startled to receive such a stark declaration, from one of geology's elder statesmen, that Moses' account of Creation was just a myth.

The following year Parkinson published *Outlines of Oryctology* – a textbook on fossils designed for students, the study of which he still called oryctology.[5] It had been advertised as being in preparation as early as 1804, but when Parkinson had learnt that someone else was preparing a similar introductory book he had decided not to continue; however the competing work never materialised.[6] In *Outlines* he reaffirmed his belief in the 'perpetual influence of a Divine Providence' while reiterating many of the points made above to Buckland. Now 67, it is remarkable how he still retained a sense of almost childlike wonder at the extraordinary events that had occurred during the Earth's ancient history, daily revealed to him through this young science of geology. In the eleven years since he had completed *Organic Remains*, Mary Anning had discovered the first fossil ichthyosaur, Edward Jenner had found the first plesiosaur and many other remarkable beasts had been revealed, although it would still be another 20 years before the term 'dinosaur' was coined. To Parkinson, these extraordinary animals were the most definite proof that 'another world' had existed in the distant past.

From his letter to Buckland, it is quite clear that Parkinson was by now firmly of the opinion that humans had arrived on Earth *after* the biblical Deluge in which they were supposed to have perished. But despite having reprimanded Buckland just a year earlier for not clearly stating this to be the case, when it came to discussing the creation of Man in *Outlines*, Parkinson was still apprehensive about revealing his true beliefs to the public. While he did state, 'The Mosaic account of this deluge has, however, been doubted, from the total absence of the fossil remains of man', he stopped short of saying unequivocally that Moses' account of Creation was 'uninformed', as he had to Buckland, suggesting instead that human remains might not have been fossilised due to unfavourable conditions. Such prevarication makes clear that when writing for the public, Parkinson was still nervous about offending their religious sensibilities – as Darwin would be more than 25 years later. Over the years, his knowledge of geology and all that it revealed about the formation of the Earth meant he had had to adapt his faith to accommodate the indisputable facts presented by fossils. But despite his promise 'to conceal no conclusion, however repugnant to popular opinion or prejudice', he did not have the temerity to pass on to his audience everything he had learnt. Nevertheless, he ends *Outlines* reiterating his belief in a world of 'indefinite duration' in which animals had come and gone in a series of successive creations, and predicts that Man 'in his turn, is destined, with the Earth he inhabits, to pass away, and be succeeded by a new heaven and a new earth'.

Within weeks of the publication of *Outlines*, Parkinson received a most welcome surprise. Back in 1800 the Company of Surgeons had been granted a Royal Charter, becoming the

Royal College of Surgeons, as it is still known today. Two years later the College established an Honorary Gold Medal to be awarded for 'liberal acts or distinguished labours, researches and discoveries eminently conducive to the improvement of natural knowledge and of the healing art'. After twenty years, they still had not found anyone of sufficient calibre to be its first recipient, so it was to Parkinson's great astonishment that a letter informed him the first Honorary Gold Medal would be awarded to 'Mr. James Parkinson of Hoxton Square',

> . . . in consideration of his useful labours for the Promotion of natural Knowledge, particularly that expressed by his splendid Work on Organic Remains – and of his liberal and valuable information, when called upon by the College, in its Research for facts relating to its scientific Designs.[7]

Accordingly, and appropriately on Parkinson's 68th birthday,[8] he was made an Honorary Member of the Royal College of Surgeons and decorated with its Gold Medal. Sir William Blizard, the College President, delivered the oration, praising Parkinson for having provided the College, whenever asked, with information about fossils, for allowing other naturalists to consult his collection, and for the general 'tenor' of his scientific life. In particular, he considered that Parkinson's work on both characterising fossils and identifying the means by which they had been altered over long periods of time would prove invaluable not only to understanding the 'physical Changes of this Globe', but also to the sciences of anatomy, physiology and chemistry. Furthermore, by inspiring others to take up the subject, his work would live on. Sir William then conferred the medal on him:

> Mr. James Parkinson: by the Authority, and in the Name, of the Royal College of Surgeons in London, I deliver to you this

> Honorary Medal . . . And may you long enjoy the sweetest Solace of life, Reflect[ing] on your useful Works.

Mr Parkinson stood up and thanked Sir William, modestly protesting that the council had overestimated the value of his work. He went on to explain how his interest in fossils had been sparked by having seen Hunter's collection almost 40 years ago, but he felt obliged to point out that solving some of the problems in palaeontology would not necessarily contribute to a better understanding of anatomy and physiology, because the fossil remains of extinct species were so different from animals living today. Having thanked everyone for their favourable opinion of his exertions, Mr Parkinson withdrew while the council finished its business.[9] Before the dinner could commence it was necessary for the council to suspend its regulation which prohibited 'any member of the College in actual practice' to dine with council members, which they accordingly enacted. It was an exceptional honour.

Once the meeting was over, Parkinson sat down to dine with Sir William and eighteen members of council, which included all the great surgeons of the day: John Abernethy, Henry Cline, Sir Astley Cooper, Sir David Dundas, who had been physician to King George III; Sir Anthony Carlisle, then Surgeon Extraordinary to George IV; as well as his predecessor Professor Thomas Chevalier and twelve others. It was a magnificent gathering.

James and his son John were indeed still 'in practice', still the parish doctors attending the local paupers of the workhouse, and struggling to cope with yet another epidemic of typhus. To do so they had introduced a new regime for infected patients, which was proving remarkably effective. On discovering an inmate of the workhouse had the disease, the person was immediately conveyed to the fever ward where they were given a bath, their hair

Sir William Blizard when he was Professor of Anatomy and Surgery to the Royal College of Surgeons.

was shaved and they were made to put on clean linen, with an upper garment of wool. They were then bled, given an emetic, and a blister was applied to the nape of the neck or 'any part in which congestion is threatened'. Clean and purged, the patient was then seated in a chair with a high back to support their head and placed 'in the current fresh air passing upwards from the windows . . . through the openings in the ceilings and roof'. They were forbidden to lie down during the day and a light diet was prescribed, wine and beef in particular being prohibited. Sometimes no food at all was allowed for several days. In addition, no visitors were allowed, which may have contributed significantly to the scheme's success.

Of the 187 patients treated this way, only five had died and two of these had been nearly 80. It was a remarkably successful trial and the Parkinsons strongly recommended the treatment to their more wealthy patients. Here, unfortunately, their success rate was not repeated, because, the Parkinsons believed, these patients were treated in their own homes and did not adhere strictly to the regime. 'Anxious friends' could not bring themselves to administer the privations required and 'the tender nurse' would inevitably provide them with 'supposed harmless' indulgences when asked. But, the Parkinsons insisted, when 'death is threatening. . . the proposed measures must be adopted and employed with unyielding resolution'.[10] Eager to promote the treatment they had discovered, father and son wrote a joint paper *On the Treatment of Infectious or Typhoid Fever** which was published in March 1824.[11] It was to be the last work James ever published.

John lived near his parents at 3 Pleasant Row, just off the Kingsland Road in Hoxton, with his new wife Eliza and their young baby, as well as two children from John's first marriage. His first wife, Diana Chapple, had died in 1820 while giving birth to their third child, John Chapple Parkinson, who only

* The two diseases, typhoid and typhus, had not yet been distinguished from each other and the terms were used interchangeably. Infectious typhoid was in fact typhus. Although many of the symptoms (and the names) of these diseases are very similar, often making diagnosis problematic, they otherwise could not be more different.

survived a few months. A year later, John married Elizabeth Caroline Tompson, who was known as Eliza. She was some fifteen years his junior and went on to bear him six more children, five of whom survived to adulthood. The Hoxton Square household was now rather diminished: the eldest daughter, Emma, had married John Dimock in 1816 and moved to Stonehouse in Gloucestershire, where the family grave can still be seen in the churchyard; and the youngest son, Henry Williams Parkinson, had gone to live in Calcutta, where he married Elizabeth Asperne in 1820. Only the youngest daughter, Mary Dale Parkinson, was still unmarried and living at home. So perhaps because accommodation in Pleasant Row was becoming rather crowded, the two families swapped houses. John and his family moved to Hoxton Square, while James and the two Marys, his wife and daughter, went to live at Pleasant Row. It must have been a considerable upheaval for James, who had lived his whole life at No. 1 Hoxton Square. Perhaps it contributed to a decline in his health, for a few days before Christmas 1824, John Parkinson had a sad duty to perform: that of writing to the parish's Board of Trustees to inform them of his father's death:

> Gentlemen,
>
> With feelings of deep regret I have to announce to you the Death of my much beloved and respected father.
>
> He was suddenly attacked on Sunday afternoon with a severe paralytic affection which deprived him of his speech and the use of his right side. Every means were employed for his recovery but they proved of no avail – he expired on Tuesday morning at half past ten o'clock.
>
> I have the honour to be, Gentlemen,
> Your most obedient servant,
> J.W.K. Parkinson[12]

James had suffered a severe stroke and died on 21 December 1824. He had amended his will just a month earlier, so had probably been unwell for a while.

The success of all his books meant he left a comfortable financial legacy for his family, making sure they would be provided for. John inherited the apothecary shop in Hoxton Square and all of the 'drugs, shop's utensils and things in and about the said premises' that a surgeon and apothecary might need. The legacy also included his father's books and the honorary medals 'presented me by the Humane Society and by the Royal College of Surgeons'. His glorious 'Cabinet of fossil Organic remains with the cases and drawers in which they are contained', plus any money owing to him from his publishers, went exclusively to his wife Mary.[13] Unfortunately she seems not to have entirely welcomed all of the bequest.

On 3 December 1825, less than a year after Parkinson had died, his son John wrote to Sir William Blizard, who at 82 was still President of the Royal College of Surgeons. John informed Sir William that he and his mother, as co-executors of Parkinson's will, were about to dispose of his father's 'collection of fossil organic remains', and he offered the College first refusal to purchase the collection for the Hunterian Museum, now housed on the College's premises. The College looked favourably on the offer and agreed to select the specimens it 'judged desirable for the College to possess'. These they would have valued 'by some competent person or persons and purchase the same at such valuation'. Not unreasonably, mother and son rejected the proposal on the grounds that they did not want individual specimens removed from the collection.

Perhaps rather affronted, the College asked John to let them know what his minimum price would be for the whole collection, but before he could discuss the matter with his mother he received a letter from her saying the collection should not be sold

John Hunter's Museum at the Royal College of Surgeons, 1853.

for less than £1,500. Maybe she anticipated the College could not, or would not, afford this high price, so she told John she intended seeking a buyer from elsewhere and therefore discussions with the College were at an end. The tone of John's letter to Sir William reporting his mother's wishes implies he regretted her decision, although he could never say so directly. At its next meeting the College board resolved 'That in consequence of such letter this Board considers all negotiation respecting the purchase of the collection is at an end'.[14] It was a sad day for posterity.

More than a year went by during which the Parkinsons tried and failed to find anyone prepared to pay £1,500 for the whole collection. Eventually, early in April 1827, advertisements appeared in the newspapers stating that Mr George Brettingham Sowerby,[15] auctioneer,

> respectfully informs the Nobility and Gentry who patronise the scientific study of Geology, that on the 9th, 10th, 11th, and 12th of the present month of April he will submit to SALE by PUBLIC AUCTION the whole of the capital collection of fossils & organic remains of the late James Parkinson Esq, author of "The Organic Remains," together with his library of Geological Books.[16]

That the collection was to be sold over four days is an indication of its very considerable size. It generated a great deal of interest and 'the Nobility and Gentry' duly assembled in Sowerby's auction rooms in Regent Street on the appointed day. George Featherstonhaugh, a British-American geologist and geographer, bought Parkinson's beautiful collection of silicified zoophytes 'for a few pounds'[17] and took them to the United States where they were subsequently lost in a museum fire. The Reverend Adam Sedgwick, then Professor of Geology at the University of Cambridge, attended each of the four days 'at his own expense' and subsequently recorded the acquisition of 'A large collection of very magnificent fossils'. He considered that his 'much larger than usual' expenditure on improving the university's collection was an opportunity 'wich [*sic*] may perhaps never again occur'.[18] Unfortunately these specimens subsequently became amalgamated into the Sedgwick Museum's collection and cannot now be identified.[19] Many other collectors bought specimens for small sums and Parkinson's lifetime collection soon became dispersed around the world.[20] His young friend, Gideon Mantell, remembered the sad occasion well:

> The matchless collection of Mr. Parkinson (the author of the 'Organic Remains of a former World'), which contained most of the specimens figured in his beautiful work, was disposed of by auction, and realised a very inadequate sum. A fragment of a Cidaris [sea urchin], in Flint, surrounded by spines, figured in Org. Remains vol. iii; and which (Mr. Parkinson informed me) cost him 20 guineas, fetched so small a sum that it was subsequently purchased of a dealer for £3. 3s.[21]

Mantell later described how Parkinson had purchased this particular fossil

> in the palmy days of the study of organic remains, before the terms Geology and Palaeontology were invented, and when a choice relic of "a former world" was cheap at any price, in the opinion of the enthusiastic collector.[22]

In fact Parkinson had purchased it in 1806 at the sale of the Leverian Museum where it had been considered one of the most valuable specimens.[23] Parkinson had also paid twenty guineas for the superb Stone Lily, a sketch of which adorns the second volume of *Organic Remains*. It was purchased by the Marquess of Northampton and Mantell estimated that such a specimen was now only worth five to ten guineas.[24] Prices had dropped considerably since Parkinson had made his collection, largely due to the fact that he had been collecting during the long wars with France when obtaining fossils, particularly from abroad, was difficult and consequently expensive. In 1806, for example, he had dropped out of bidding for a fossil echinoderm when

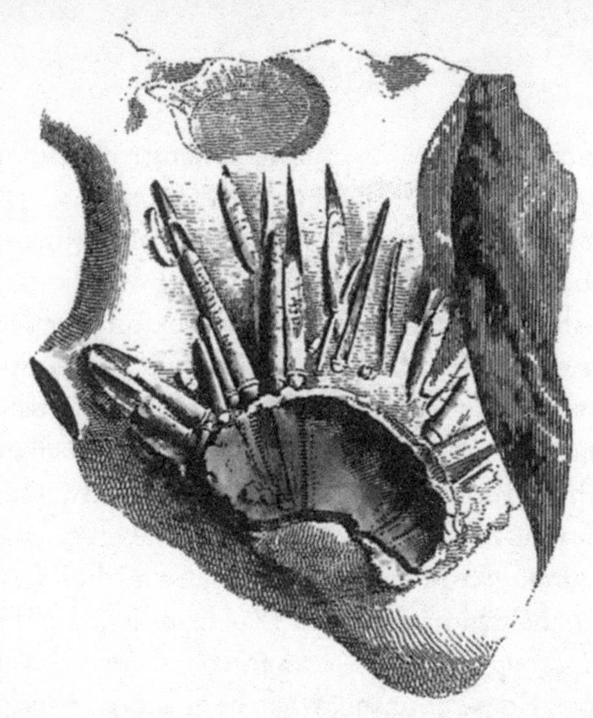

The 'exquisite' fossilised sea urchin, Cidaris, that was in Parkinson's collection. The interior of the fossil was a light pink colour.

the price reached fifteen guineas (£15 15s); it eventually sold for £15 17s 6d. When sold again in 1828 it fetched only £3 15s.[25]

One of the more spectacular fossils that Parkinson owned, used to illustrate the frontispiece to his second volume of *Organic Remains*, was a silicified cup-shaped sponge, initially called *Chenendopora parkinsoni*, but now called *Chenendopora michelinii.*[26] This specimen is one of the few fossils known today to have been in his collection. If you ask very nicely, it, along with half a dozen others, can be seen at the Natural History Museum in London.

We do not know whether Mary Parkinson realised her hoped-for £1,500, because no catalogue from the sale has ever been found, but from Mantell's comments it seems unlikely.[27] Whatever the amount, she had to wait for her money: a letter written almost a year after the auction by John Parkinson, on behalf of his mother, requested that George Sowerby 'settle the account as soon as possible as circumstances of an urgent nature depend upon it'.[28] We can but wonder what those circumstances were.

After his father's death, John continued for another nine years as parish doctor and medical attendant to the workhouse and poor of the parish, alongside managing the Hoxton practice. Five of those years were without anyone to help him; the workload must have been tremendous and he complained to the Trustees of the Parish that parish patients were taking up far too much of his time, visiting as he did some 70 to 80 sick paupers a day. This left him insufficient time to attend to his private midwifery cases, which were his main source of income. Although a medical assistant was appointed to help him in 1829, in 1833 he resigned and moved to Islington to set up in practice as a surgeon; there he was joined by his son, James Keys Parkinson, when he qualified as an apothecary the following year. Following John's resignation, no fewer than six practitioners were appointed as parish doctors in Hoxton to carry out the work formerly done by two men.

John William Keys Parkinson died aged 53, just five years after he moved to Islington, on 5 April 1838.[29] He died on the same day that his mother was buried, Mary having died eight days earlier.[30] Both death certificates state they died from 'fever', suggesting they were victims of the typhus epidemic that stalked London and so many other cities that year. The epidemic had started in July 1837 and the deaths in England and Wales during the ensuing eighteen months numbered almost 30,000, a quarter of which were in London. The distress of the poor during this terrible epidemic alerted the public to the deplorable conditions of filth and overcrowding in which so many people lived, causing the great health reformer, Edwin Chadwick, to produce a pioneering report which recommended the urgent need for improved living conditions and medical care for the poor.[31]

Despite starting his career as a lowly apothecary, James Parkinson rose to become a significant figure in London's medical fraternity. The 1816 *Biographical Dictionary of Living Authors* called him an 'ingenious practitioner and naturalist',[32] and testimonials in medical publications verify that he was also held in high esteem by the general public. One reader of *The Hospital Pupil* considered the author 'treated his various subjects in so judicious a manner, and has expressed himself with so much perspicuity, that no medical gentleman ought to commence his routine of studies until he has carefully perused these letters'. *Medical Admonitions* provided 'uncommon satisfaction with the design, as well as with the manner in which it is executed', while parents were 'duty bound' to put *Dangerous Sports* 'into the hands of the rising generation'.[33] His *Essay on the Shaking Palsy* was received in a similarly favourable manner, but in its day it was just another pamphlet in his long list of publications.

Parkinson's *Essay* provided the first clinical description of the disorder now known throughout the world by his name. Although successive generations of neurologists have refined and embellished his seminal account, diagnosis of the disease remains dependent on recognising the symptoms being presented, just as Parkinson recognised them 200 years ago.

This good man would probably be surprised to know the disease now bears his name, and astounded to discover that his birthday, 11 April, has been designated World Parkinson's Disease Day. In 1980 a red and white tulip was named in his honour by a Dutch horticulturalist with Parkinson's disease, and it has become the symbol for Parkinson's disease societies around the world.[34] But despite all these accolades, of which he would no doubt be very proud, I suspect James Parkinson would rather be remembered for a different publication.

Mary Parkinson's will testifies that she did keep some of her husband's fossil collection (or were they just pieces left over from the auction that did not sell?) since she left 'half the fossils and minerals' to each of her daughters, Emma and Mary. Emma had coloured many of the plates in *Organic Remains*, and today it is those exquisitely illustrated plates that we particularly cherish; originals can still be purchased for considerable sums. The importance of these illustrations lived on for decades after Parkinson's death, and in 1850 Mantell reprinted them in his *Pictorial Atlas of Fossil Remains*, reinterpreting the fossils they represented in the light of more recent knowledge. In the Preface to that work, Mantell paid homage to his mentor:

> I gladly avail myself of this opportunity to make a passing allusion to the excellent and accomplished author, Mr. Parkinson. I had the pleasure and privilege of his acquaintance in my youth, immediately after the publication of the third volume of his valuable work.
>
> The publication of Mr. Parkinson's *Organic Remains of a Former World*, at the commencement of the present century, must be regarded as a memorable event in the history of British Palaeontology: it was the first attempt to give a familiar and scientific account of the fossil relics of animals and plants, accompanied by figures of the specimens described.
>
> At that time, the terms Geology and Palaeontology were unknown; all the sedimentary strata have since been accurately defined and arranged, and names assigned to the respective systems or formations; while the so-called *Diluvial Epoch*, which Mr. Parkinson, and even Baron Cuvier, considered as established by incontrovertible physical evidence, has been expunged from the chronology of geology.[35]

This tribute is testimony to the high regard in which the scientific community held Parkinson's work on fossils. Indeed, his *Organic Remains of a Former World* and its successor, *Outlines of Oryctology*, were so successful that in later life he was often referred to as 'Mr Parkinson the Oryctologist'. *Organic Remains* remained in print for some 40 years after publication of the first volume,[36] during which time much new evidence came to light regarding fossils and geology. The volumes must have looked rather out of date well before the 1840s, since Parkinson made no changes to them whatsoever. It can only be assumed that the publishers were still selling them in sufficient numbers that it made their quite considerable financial outlay worthwhile; it also reflects the immense popularity of the work. The frequency with which the work was cited, at least until the 1830s, shows that it also provided the serious researcher with a much-needed reference tool.[37]

Parkinson's achievements were also recognised internationally. In Moscow he was made an honorary member of the Imperial Society of Naturalists; the German naturalist Heinrich Georg Bronn named a crinoid, the 'Pear encrinite', *Apiocrinus parkinsoni*, in his honour; and the fossil fruits of the Malucca palm, which Parkinson had thought were fossilised cocoa beans, were called *Nipadites parkinsoni* by the French chemist Alexandre Brongniart.[38] In England the renowned naturalist James Sowerby[39] named the Jurassic ammonite *Parkinsonia parkinsoni* after him, and Gideon Mantell remembered his friend by dedicating a species of gastropod *Rostellaria parkinsoni* to him.[40]

The high esteem in which his work was held was also evident when the Royal College of Surgeons awarded him its Honorary Gold Medal.[41] In Blizard's oration, no mention was made of Parkinson's extensive medical publications, his work on chemistry, or even his *Essay on the Shaking Palsy*; instead, it was his outstanding contribution to the field of palaeontology – his favourite science – for which he was honoured. I am sure it is this legacy for which he would most want to be remembered today. His diligent research over many decades revealed a world hitherto unknown; one that was littered with 'wrecks of an early Creation' that were 'entombed in the bowels of the earth'; a world populated with 'hyenas the size of bears' and the 'relics of a tribe of enormous marine animals . . . possessing the blended structure of fish and lizards',[42] all of which both enthralled and terrified his readers. When had these creatures lived? How had they arrived on this planet? And how come they had invaded our shores when their relatives now lived in tropical climates thousands of miles away? These were all problems he grappled with.

Unable to find definitive answers himself, Parkinson offered these questions to his public in the hope that 'philosophical Enquirers' of the future would be stimulated to answer them, in the same way that he hoped medical men of the future would be 'excited to extend their researches to this malady' when he published his *Essay on the Shaking Palsy*. This farsighted, questioning approach meant that Parkinson left us with a remarkable scientific and medical legacy. Indeed, as Blizard said when awarding him the Gold Medal, the light Parkinson had shone on science would benefit mankind 'until the end of time'.

The Jurassic ammonite *Parkinsonia parkinsoni*.

EPILOGUE

A fragment of DNA

DNA neither cares nor knows. DNA just is. And we dance to its music.

Richard Dawkins, 1995
River Out of Eden

LIKE GENERATIONS of first sons before him, James Warren Parkinson was expected to become a surgeon.[1] His father, James Keys Parkinson, was the son of John William Keys Parkinson and the grandson of James Parkinson; James Warren Parkinson was therefore James Parkinson's great-grandson. James Keys was granted the apothecaries' Certificate of Qualification on 29 March 1834,[2] and when his father died during the typhus epidemic of 1838, he took over the business. Tragically, James Keys died of tuberculosis on 24 February 1849, aged only 36. He left behind his wife Sarah Anne and their four surviving children (of seven born): James Warren aged twelve, Emma aged six, Rosina aged two, and Margaret, a baby of nine months.

Without her husband's income, James Keys Parkinson's wife, Sarah Anne (née Warren), would have struggled financially. Indeed, two years after James Keys died, the 1851 census

James PARKINSON (1755–1824) married Mary DALE (1757–1838), both born, married (1781) and died in Shoreditch, London.

John William Keys PARKINSON (1785–1838) (eldest surviving of 7 children), born and died Shoreditch, married (1811, Shoreditch) Diana CHAPPLE (1787–1820), born Sevenoaks, Kent.

James Keys PARKINSON (1812–1849) (eldest of 3 children), born and died Shoreditch, married (1837, Milverton, Somerset) Sarah Anne WARREN (1819–1892), born Milverton, died France.[3]

James Warren PARKINSON (1837–1900) (eldest of 7 children), born Shoreditch, married (1874, Dunedin, New Zealand) Catherine COOKE (1850–1932) born Stroud, Gloucestershire, both died Dunedin.

Sarah Anne May PARKINSON (1875–1960) (eldest of 11 children) married (1902, Dunedin) Charles WARRINGTON (1880–1923), both born and died Dunedin.

Sophia May WARRINGTON (1904–1996) (2nd of 5 children) married (1930, Dunedin) Alfred Gilbert REYNOLDS (1901–1970), both born and died Dunedin.

Robert Alfred REYNOLDS married Judith Ellen THOMSON (now living in Dunedin)

John Charles REYNOLDS (born Dunedin) (now living in Christchurch, New Zealand)

Reuben and Flynn REYNOLDS (now living in Christchurch)

Table of James Parkinson's descendants discussed here.

Is this the face of James Parkinson?
James Keys Parkinson, James's grandson.

records that Sarah Anne had become a governess; James Warren Parkinson and his sister Emma had been placed in a school, but Rosina was in an orphanage. There is no mention of baby Margaret, although she is likely to have been in an orphanage as well. In the mid-nineteenth century people saw large-scale institutions such as orphanages as the best way of providing for children whose families could not support them, but it must have been unbearably hard for Sarah to part with them. She did manage to reunite the family eventually, as evidenced by the photograph below, and it seems most likely this happened when she married again, although that was not until 1866, by which time her children were young adults. Her new husband, John Spurway, was a gentleman of some means who had been a lieutenant in the Navy. He lived in Milverton in Somerset, where Sarah herself was from originally; he too was a widower with grown-up children.

Sarah Anne Parkinson (née Warren), the wife of James Keys Parkinson, with their children: James Warren and Emma (back), Rosina (front left) and Margaret.

The family story, passed down through the generations, is that James Keys intended his eldest son, James Warren, to become a surgeon and, given the family's background, this is no doubt the case. James Warren Parkinson's great-grandfather, James Parkinson, had left the shop in Hoxton Square to his eldest surviving son, John William Keys Parkinson. In turn, the will of John William Keys stipulates that 'my son James Keys Parkinson' should inherit 'every article or thing appertaining to the practice of the art of surgery and pharmacy and all my books'. James Keys undoubtedly did the same for James Warren. There was only one problem with this plan – James Warren did not want to be a surgeon because the sight of blood made him feel ill.

His desire to break with the centuries-old family tradition must have upset his mother considerably – and James Parkinson would have turned in his grave – but nevertheless, the story goes, James Warren Parkinson was allowed to give up surgery and instead was bought a commission in the Army.[4] He was then sent out to New Zealand as an *aide-de-camp* to a general where he acted as an interpreter during the Māori wars, a series of armed conflicts between the New Zealand Government and the indigenous population that took place between 1845 and 1872. James Warren apparently had a flair for languages and already spoke four by the time he arrived in New Zealand, so it did not take him long to learn Māori. Being able to speak the local language, without his superiors being able to understand what he was saying, may have led them to think he was not always being entirely truthful, for at one point he was apparently sent back to England for 'fraternising with the enemy'.

But James Warren had taken a liking to New Zealand and it was not long before he was back. At his wedding in Dunedin in 1874 he gave his address as Auckland, so presumably he had been living in New Zealand for some time before then. His wedding certificate also states that he was a widower, but it is not known if he had children by his first wife, or who she was. What we do know is that on coming back to New Zealand, on a sailing ship called the *May Queen*, James Warren had met his second wife, Catherine Cooke.

The *May Queen* was a small barque and popular with passengers.[5] She left Gravesend around 6pm on 19 August 1874 with 70 passengers on board (twenty in saloon class, sixteen in second cabin and 34 in steerage), as well as 30 crew under the command of a Captain R. Tatchell. The voyage to Otago on New Zealand's South Island took a fairly quick 84 days and was relatively uneventful, despite the *May Queen* being 'humbugged by singularly variable weather'. This meant that although they had travelled 916 miles by 21 September, only 420 of them had been in the right direction. When passing through the tropics a

The sailing ship *May Queen* in Nelson circa 1880.

note in the *May Queen Weekly News*, an onboard news sheet that James Warren Parkinson co-edited, requested passengers to 'preserve their equanimity' while the calm weather caused the sails to 'hang like boards against the stately masts and the tropical sun poured down its heat upon the deck'.[6]

The only major incident during the whole voyage was the death of one of the saloon passengers: after a few days' illness 'Henry Chase Edmonds aged 34 died of acute mania and debility'. Shortly before he became ill, Edmonds had sent this touching correspondence to the *May Queen Weekly News*: 'The sunrise was magnificent this morning, and this evening I saw the most splendid sunset I have ever beheld. The moon and Venus rising in close proximity to each other were exceedingly grand; and I think we ought to thank God for all his goodness to us.' His remains were consigned to the deep with befitting ceremony on the day that he died. His poor widow had no choice but to remain on board.[7]

Catherine Cooke appears to have been travelling on the *May Queen* on her own, leaving both her parents and her several brothers and sisters back home in Stroud, Gloucestershire. The previous year the New Zealand Government had waived the £5 adult fare in order to encourage immigration, and Catherine was taking advantage of this attractive opportunity. Towards the end of the Māori wars, New Zealand had entered a period of depression, the wars having created a negative impression of the place overseas which had discouraged immigrants. The Government believed that in order to make the nation prosperous, large-scale immigration was needed to create an economic boom. In particular, immigrant labour was badly needed to build railways and roads, and to work on the land, so the Government encouraged them by financially assisting their passages and settling them on land purchased or confiscated from the Māori. In addition, existing New Zealand residents could nominate friends and relatives to come and join them. Newspaper advertisements and posters in Britain called for married agricultural labourers and single female domestic servants to join the exodus, provided they were 'sober, industrious, of good moral character, of sound mind and in good health'. As a consequence, the great migration to New Zealand began in 1871. Three years later, in 1874, the year Catherine Cooke came out, some 38,000 immigrants set out to start a new life there, three-quarters of them coming from England. That year the number of women emigrating outstripped the men.[8]

It is possible that Catherine, then just 24, was going out to join her elder sister, Ellen, and Ellen's husband, Duncan MacArthur: it is known that both of them were buried in New Zealand. At 37, James Warren was thirteen years Catherine's senior, but the two would have had plenty of opportunity to get to know one another. For although a few amusements such as blind man's buff, quoits, singing and dancing were provided on board, there was little else to do on the three-month voyage other than walk the decks and talk. Eventually the *May Queen* docked at Port Chalmers in Otago Harbour, Dunedin, on Friday 13 November 1874. Three days later Catherine and James were married by the Reverend Dr Stuart.

That afternoon, Captain Tatchell and Mr Walker, his chief officer, were entertained at a banquet provided by the passengers, at the Provincial Hotel, Port Chalmers: 'Twenty ladies and gentlemen sat down to a capital spread . . . The entertainment originated in a hearty appreciation on the part of the passengers of the undagging efforts of Captain Tatchell and his officers to promote their comfort and amusement during the voyage and was arranged by Messrs. Parkinson, Thacker, Hamilton and Duder, the committee appointed for the purpose.' After loyal toasts were given, the Chairman of this committee thanked the Captain on behalf of himself and fellow passengers 'for the unvarying kindness and attention he had lavished upon them during the passage'. He believed that Captain Tatchell was everything that the commander of a crack ship like the *May Queen* ought to be, and concluded by designating the *May Queen* as the queen of ships and the Captain as the king of commanders. James Warren Parkinson, Honorary Secretary of the committee, then read an address to Captain Tatchell that bore the endorsing signatures of 51 passengers. The address, together with a handsome locket of locally manufactured Otago gold, was then presented to the Captain. Clearly everyone had had an excellent time on board, thanks to the efforts of the ebullient Captain Tatchell and his crew. The celebration also served as a wedding reception for the newlyweds, since it was attended by many of the couple's friends made on the voyage. It was certainly an auspicious start to their new life together.

Although the Otago gold rush of the 1860s was well over by the time the Parkinsons settled there, the new colonial settlement of Dunedin had reaped the benefits of the boom. It rapidly commercialised and expanded, quickly growing to be New Zealand's largest city, a position it retained until about 1900. Consequently, there were plenty of opportunities for a young man keen to make his way in the world and James Warren Parkinson took full advantage of them, his family flourishing as a consequence. The photos of his children all show well-dressed and prosperous-looking young men and women. He was also of good moral standing, becoming District Deputy of the Independent Order of Templars in Dunedin, a group modelled on Freemasonry with a mission to enlighten people around the world about a life free from alcohol.

Ten months after their wedding, in September 1875, Catherine gave birth to their first child, Sarah Anne May, and further children were born in each of the following two Septembers. Over the following twelve years they had another eight children, making eleven in total, nine of whom lived to have children of their own.

Today a large number of James Parkinson's descendants still live in Dunedin and Christchurch and I managed to locate one, John Reynolds, through his excellent genealogy website, which includes information on 119 of Parkinson's descendants.[9] When I contacted John he introduced me to his great-grandfather, James Warren Parkinson (James Parkinson's great-grandson), and told me what the family knew of his story. As it happens, one of my own sons now lives in New Zealand too so, anxious to meet one of Parkinson's descendants in the flesh, on my next visit I arranged to meet John Reynolds in Christchurch where he now lives, although his parents are still in Dunedin.

Catherine Parkinson (née Cooke) with her grandson,
Frank Alfred Warrington (1905–1962).

Frank Warren Parkinson (1880–1953) (left) and his younger brother Henry (1882–1946).

Three of James Warren Parkinson's daughters: Sarah (1875–1960), Margaret (1886–1925), centre, and Susan Catherine (1878–1952).

I spent the morning touring the centre of Christchurch, which had been so devastated by the massive earthquakes of September 2010 and February 2011. Despite the damage, still so visible almost four years later, and the fact that so little restoration work had begun, the place felt vibrant and exciting. Shipping containers had been turned into a shopping mall, a cathedral had been built out of cardboard, art in all its forms was visible everywhere, and an intensely moving exhibition, 'Quake City', showed what living through the earthquake had been like. I shall always remember the film of a young man trying to keep his balance on his bicycle as the ground shook beneath him.

When the magnitude 6.3 earthquake struck Christchurch on 22 February 2011, at 12.51 in the afternoon, John Reynolds, a computer programmer, was working in the Bank of New Zealand building in Cathedral Square, right in the city centre. He was in the kitchen on the third floor making some lunch. When the shaking started John looked up to see one of many buildings topple – much of Christchurch's infrastructure had already been weakened by the magnitude 7.1 earthquake of the previous September – and by the time he got outside, the famous cathedral spire was already down, leaving only the lower half of the tower standing. Through the clouds of dust he watched with horror as the Canterbury Television building then collapsed, killing 115 people. Understandably, his first thought was for his own children; Reuben was at home with John's wife and John managed to establish that they were safe, but Flynn was at school some distance away. The bridge John had to cross to get there had come down in the quake and he remembered having to wade through sewage to reach the school and his son, who was safe. John's memories of that day were still very raw.

As I listened to him talk, I wondered what James Parkinson would have made of it all. Earthquakes were still unexplained in his time, but I felt certain he would have been both interested

James Parkinson's great, great, great, great, great grandson
John Reynolds in Christchurch, December 2014.

in and amazed by what we know about them today. New Zealand sits astride the boundary between the Australian and the Pacific tectonic plates – huge pieces of the Earth's crust that are constantly moving around the globe. Along the eastern side of the North Island, the Pacific Plate is being subducted (sliding down) beneath the Australian Plate, and this is what causes the volcanic activity in the North Island and the majority of the earthquakes. About 20,000 earthquakes are recorded in New Zealand each year, although only about 200 of these are strong enough to be felt. In November 2016 another massive quake struck the South Island and was felt the entire length of the country. At magnitude 7.8 it was more than ten times stronger than the February 2011 quake in Christchurch and at two minutes in duration it was one of the longest ever recorded anywhere,[10] but although it caused widespread devastation there was fortunately little loss of life.

Down the length of the South Island, the tectonic plates are sliding past each other along the Alpine Fault. In the 1940s the geologist Harold Wellman realised that large areas of rocks in Nelson, which is on the north-west corner of the South Island, matched rocks around Otago on the south-east coast. He did this by comparing the fossils and other features in the rocks, and it led him to realise that the eastern side of the country has moved over 300 miles south along the Alpine Fault during the past 20 million years. Parkinson would have been delighted to know it was fossils that provided the key to unlocking New Zealand's geological history.

John Reynolds didn't look anything like I imagined James Parkinson had looked – he was tall and broad-shouldered, quite the opposite of Gideon Mantell's description of Parkinson – but he was clearly proud to be descended from him and I in turn was delighted to have met a direct descendant.[11] After we had talked for several hours, we shook hands as we got up to leave. In doing so John unwittingly transferred a fragment of James Parkinson's DNA from his hand to mine; I held it tight, reaching out to James Parkinson across 200 years.[12]

A FEW WORDS OF THANKS

Many years ago the School of Earth Sciences at the University of Bristol accepted me as a mature student, a decision which changed my life. I am now privileged to be one of their Honorary Research Fellows. I cannot thank them enough for allowing me this position, which crucially provides access to libraries, both online and off. It would have been so much harder to have done the research behind this book had I not had this support. Michael Richardson, Senior Librarian in the University Library's Special Collections, and champion of their magnificent Eyles Collection (possibly the best collection of historic geology books in the country), is a fount of knowledge and always ready to help at the shortest notice. There are also several excellent people in the IT department, particularly Catherine, who patiently got me out of a hole every time I fell into one with my computer.

It is inevitable that when doing research of this nature over several years that a number of people will have helped in some way. Had it not been for Mike Robins's brave stance against animal rights activists at a time when they were vocal and threatening, I might never have thought about writing a biography of James Parkinson at all. Thank you, Mike, for allowing me to tell your story; I shall always be in your debt. Tipu Aziz, Professor of Neurosurgery at the John Radcliffe Hospital in Oxford, performed the deep brain stimulation operation on Mike, and both he and Mike reviewed relevant sections, providing invaluable insights and feedback. Honorary Professor Neil Williams at the University of Bristol was also very helpful regarding skin cells and Parkinson's DNA.

Professor Hugh Torrens, one of our most formidable historians of science, generously gave me the file of material he had been collecting on James Parkinson from as far back as the 1970s. Including, as it did, copies of some of the extant letters written to and from Parkinson, it was a tremendous starting point for my research. Wendy Cawthorne at the Geological Society of London's library almost always had an answer to my questions and, if not, she knew someone who did. Susan Snell, Archivist and Records Manager at the Library and Museum of Freemasonry, was someone Wendy directed me to. Susan discovered that Parkinson had been a Freemason and put me in touch with Barry Evans, Secretary to the Moira Lodge where Parkinson had been a member, who kindly allowed me to view the minute books for that period. The Royal College of Surgeons' minute books record the Gold Medal award to Parkinson as well as John Parkinson's discussions with the College regarding the disposal of Parkinson's fossil collection, so many thanks to Simon Chaplin, then Senior Curator, and the several librarians I have met there over the years, for providing access to this material. Tom Sharpe, formerly Curator of Palaeontology and Archives at the National Museum of Wales, drew my attention to the fact that they hold the letters of George Brettingham Sowerby I who organised the auction of Parkinson's fossil collection; thank you Tom.

Christopher Goetz, Professor of Neurology at Rush University Medical Centre in the USA is also an historian of medical science, particularly in regard to Jean-Martin Charcot who first named the shaking palsy 'Parkinson's disease'. He very kindly agreed to read the relevant chapters of this book and shared them with his colleague, Gian Pal, MD. Their perspicacious remarks allow me to rest assured that my medical descriptions of the cause and remedies for Parkinson's disease are correct – unless I have made subsequent errors. Several other people read chapters at different stages of the book's evolution and I thank them all for their time and thoughtful comments: Simon Mitton, Gareth Williams, Jenifer Roberts, Simon Knell and particularly Wendy Moore, whose wonderful book on John Hunter, *The Knife Man*, had long been an inspiration.

I spent quite a lot of time trying to track down some of the fossils from Parkinson's collection without, unfortunately, a great deal of success. Nevertheless, I had a very enjoyable day at the Natural History Museum where I did manage to find and photograph about a dozen known to have belonged to him. Peter Tandy, Jon Todd, Giles Miller, Jill Darrell and especially Consuelo Sendino were very patient in helping me locate them. Matt Riley at the Sedgwick Museum in Cambridge also tried to locate Parkinson's fossils there.

Librarians who have been unfailingly accommodating include those at the Wellcome Library where I viewed the various minute books of the Royal Jennerian Society, which revealed that Parkinson had worked with Edward Jenner; and University College London Archives which holds a vast collection of papers from George Greenough, an inveterate correspondent but sadly not with Parkinson. I would also like to thank the Wellcome Library for their enlightened policy of allowing their digitised historical images to be used free of charge. I could have spent much longer looking at the fascinating minutes of Guy's Hospital Physical Society at Kings College London Special Collections, and Gideon Mantell's correspondence in the Alexander Turnbull Library in Wellington, New Zealand. In the British Library I located the papers of Francis Place, so important for understanding the radical movement of the 1790s, as well as rare copies of Parkinson's works and other books not available online; the National Archives at Kew hold a large number of papers relating to the London Corresponding Society.

The detailed family tree produced by John Reynolds and

other members of his family, encompassing nine generations from John and Mary Sedgwick (James Parkinson's parents) down to John Reynolds's children, is a very impressive piece of work and has proved invaluable.

Parkinson left little in the way of personal correspondence, so in order to gain an insight into the man and his mind I was largely reliant on his published works, of which there are many, as well as articles and reviews that others wrote about his works at the time. To this end, nothing proved more useful than Google Books, where much of this material is free to download. Like everything indispensable in life, you wonder how you ever managed without it. So to all those nameless people whose fingers I occasionally see on scanned pages, as well as everyone else who has helped but not been mentioned, I raise my glass in thanks.

Cherry Lewis

NOTES AND REFERENCES

Prologue: A hole in the head

1. The revival of neurosurgery for movement disorders started in the mid-1980s. In 1987, Alim-Louis Benabid and Pierre Pollak, two French neurologists at the University Hospital of Grenoble, developed deep brain stimulation for patients suffering from Parkinson's disease (Pollak & Krack, 2007). In 2014 Benabid won the prestigious Breakthrough Prize in Life Sciences for his pioneering work in this field.
2. A video of Mike Robins controlling his Parkinson's disease symptoms can be viewed on YouTube: https://www.youtube.com/watch?v=h8tWlYv1Ykc

Chapter 1. Living and bleeding in London

1. Wolfgang Amadeus Mozart (1756–1791), Marie Antoinette (1755–1793), Joseph Priestley (1733–1804), Humphry Davy (1778–1829), William Hunter (1718–1783) John Hunter (1728–1793), Edward Jenner (1749–1823), William Wordsworth (1770–1850), William Turner (1775–1851) and William Smith (1769–1839).
2. Previous biographies of Parkinson state that his younger brother, William, lived until he was 21 (i.e.1761–1782). This is incorrect. He has been confused with a William Parkinson from the Kingsland Road, Hoxton, who was buried on 14 April 1782, but this child was only fourteen days old when buried. James's brother William was baptised on 5 March 1761 and buried on 29 August 1766, aged five.
3. The naming convention for girls stated that the first daughter was called after the maternal grandmother, the second daughter after the paternal grandmother and the third daughter after the mother. James's eldest sister was called Elizabeth so, if convention had been followed, we can assume Elizabeth to have been the name of his maternal grandmother. (Elizabeth died at fifteen months in 1754 but there were no qualms in the eighteenth century about reusing the name of a dead child for a subsequent birth, so when another daughter was born in 1756, she too was called Elizabeth.) The next daughter was Margaret Townley – probably giving us the maiden name of his paternal grandmother. This leaves us with the last daughter, Mary Sedgwick, which, according to convention, should have been her mother's name. An online search revealed two Mary Sedgwicks, both born in 1721. The parents of the one baptised on 16 April 1721 in Rotherhithe, Surrey, were called John and Elizabeth, which accords

well with Parkinson's parents naming their first daughter Elizabeth (after the maternal grandmother). The other Mary Sedgwick was baptised 16 March 1721 in Rothwell, Yorkshire. We do not have her mother's name, but her father, a butcher, was called Leon Sedgwick. According to the naming convention for boys, the second son (James Parkinson's younger brother, William) should have been called after the maternal grandfather, but there is no record at all of anyone named Leon in the Parkinson family. On balance it seems most likely that the first Mary Sedgwick was James Parkinson's mother.

4. James Parkinson's father, John, died in January 1784 aged 59, which makes it statistically more likely that his birth year was 1724, rather than 1725. Since the family seemed to be following convention for naming their daughters, it is likely they would do the same for their sons, thus the first son would have been named after the paternal grandfather, and the second son after the maternal grandfather. After an extensive search, no John Parkinson born 1724–5 with a father called James has emerged, so the search continues. The absence of birth data for John Parkinson might also explain why we cannot find a record of his marriage.
5. Stott, 2014.
6. Gideon Mantell (1790–1852)
7. Mantell, 1850, p. 13.
8. Torrens, 2008.
9. Originally Roman, the old City Gate was pulled down in 1760. The section of road to the north of the original gate, which in Parkinson's time was known as Bishopsgate Street Without, marks the start of the Roman road, Ermine Street, also known as the 'Old North Road'.
10. Miele, 1993, p. 4.
11. Porter, 2000, p. 126.
12. Colman and Garrick, 1766, p. 12.
13. Miele, 1993, p. 10.
14. Designed by George Dance, favourite pupil of Christopher Wren, St Leonard's was rebuilt in 1736 after the old medieval church fell down while the congregation was at prayer. In 1817 it became the first church to be lit by gaslight, 'an innovation most profane' (Bradley, 1914, p. 27).
15. Personal communication from the Reverend Paul Turp, Vicar of St Leonard's, 2011.
16. Rowntree, 1912.
17. Smollett, 1771, p. 363.
18. *Medical Topography of London*, 1814
19. Grosley, 1772, p. 44.
20. Simond, 1815, p. 36.
21. Parkinson, 1800a, p. 29. Simon Stott has done a considerable amount of research into where James was educated but at the time of writing has not located the school: https://searching4james.wordpress.com/page/2/

22. Cockburn, King and McDonnell, 1969.
23. Whitehouse, 2011.
24. William Godwin (1756–1836).
25. Davis, 1995, p. 189.
26. Anon. (Parkinson), 1795a, p. 2.
27. Anon., 1814, p. xx.
28. Parkinson, 1800a, p. 32.
29. Ibid. p. 115.
30. Parkinson, 1799, p. 48.
31. Parkinson started at the London Hospital on 20 Feb 1776.
32. Roberts, 1997, p. 10.
33. 'London Hospitals and Dispensaries', *The Times*, 30 January 1869.
34. Lawrence, 1996, p. 58.
35. William Blizard (1743–1835).
36. Macilwain, 1856, p. 27.
37. Parkinson, 1799, p. 451.
38. Maddocks and Blizard, 1783.
39. Parkinson, 1800a, p. 37.
40. Benjamin Gooch (d. 1780) quoted in C.N. Morgan, 1968. 'Surgery and Surgeons in 18th-Century London', *Annals of The Royal College of Surgeons of England*. 42(1): p. 26 (1–37).
41. Parkinson, 1800a, p. 85.
42. Ibid., p. 83.
43. The number of criminal acts incurring the death penalty rose from 160 in 1750 to 288 in 1815.
44. Virdi-Dhesi, 2010.
45. Power, 1886, p. 275ff.
46. Berkowitz, 2013.
47. Parkinson, 1800a, pp. 77–9.
48. Parkinson, 1800a, pp. 66–7.

Chapter 2. The hanged man

1. Parkinson, 1774, p. 161–2.
2. Rev. Caleb Fleming (1698–1779).
3. American War of Independence: 1775–1783.
4. Fleming, 1773, p. 3.
5. Knapp and Baldwin, 1824, p. 44.
6. Parkinson, 1799, p. 390.
7. Ibid., p. 392.
8. John Parkinson became one of the Humane Society's first directors. He also enrolled as a volunteer medical assistant – a practitioner willing to be called on at any time to treat victims of drowning, hanging or suffocation, which is why

he was called to resuscitate Bryan Maxey. Each medical assistant was responsible for a locality related to a section of a watercourse, since the most common reason for being called out was to attend cases of drowning. The Thames had the largest number of assistants assigned to it, but John Parkinson's watercourse was a section of the New River, an artificial channel built early in the seventeenth century to convey water to the city from springs in Hertfordshire. The channel ran southward along a winding course of 40 miles, to a reservoir near Sadler's Wells, and from there water was supplied to a large area of the city. The water in the canal was rarely more than four or five feet deep, but there were occasional drowning accidents involving children and invalids. When John Parkinson died, James took on his role as a medical assistant to the Humane Society.

9. Parkinson, 1789.
10. Cherington, 2004, p. 977.
11. Franklin's experiments were documented by Joseph Priestley in his 1767 account: *The History and Present State of Electricity, with original experiments*. J. Dodsley, J. Johnson, B. Davenport and T. Cadell, London.
12. Anon., 1818, p. 249.
13. Royal Humane Society online identity statement: http://www.aim25.ac.uk/cgi-bin/vcdf/detail?coll_id=7228&inst_id=79
14. *The Times*, 26 April 1800.

Chapter 3. Fear of the knife

1. Samuel Dale (1659–1739) was a famous botanist and keen geologist, being the first to describe the fossils and strata of the cliffs at Harwich. He bequeathed his books on botany and his herbarium to the Society of Apothecaries.
2. Macfarlane, 1986, quoted on p. 124.
3. Sharpe, 1998, p. 11.
4. Corner and Goodwin, 1953, p. 363.
5. Ascites comes from the Greek word *askites*, which means 'baglike'.
6. Power, 1886, p. 376.
7. Ford, 1987, p 172.
8. Kaufman, 2001, p. 3.
9. Royal College of Surgeons Archives, Transcripts of notes of John Hunter's lectures made by other scribes, 1828–1831. MS0007/1/7/5 (Series) 3.
10. Moore, 2005, pp. 433–4.
11. White, 1880, p. 373.
12. Parkinson, 1800a, p. 72.
13. Another reason for John Parkinson wanting to publish *Hunterian Reminiscences* was that his father's notes represented a rare example of the contents of Hunter's lectures. Some thirty years after Hunter's death in 1793, all his lecture notes and manuscripts were lost in a fire. During that time Hunter's

brother-in-law, Sir Everard Home, had stepped into Hunter's shoes and carved a glittering career for himself based, many suspected, on work done by Hunter that Home had plagiarised from Hunter's manuscripts while they were in his possession. Home admitted that the fire had started while he was burning the manuscripts, supposedly fulfilling Hunter's dying wishes, but if that were true, why had Home waited thirty years before carrying out Hunter's request? As John Parkinson made clear in the Preface, few seem to have believed Home's explanation for the fire: '[The Editor] trusts the work will, in some measure, supply the loss which surgical science has sustained by the destruction (by accident it is to be hoped) of the original notes from which Mr. Hunter lectured.'

14. Parkinson, 1833, p. xi.
15. Parkinson, 1833, p. 40.
16. Parkinson, 1833, p. 37.
17. Parkinson, 1799, p. 469.
18. Fanny Burney (1752–1840).
19. Both quotes from Sabor and Troide, 2001, p. 439ff.
20. Mason Good, 1795, p. 146.
21. Porter and Porter, 1989, p. 126–7.
22. Sigsworth and Swan, 1982.
23. Porter and Porter, 1989, p. 54.
24. According to the *Oxford English Dictionary*, the word 'pharmacist' was first used in a publication in England in 1834 in a novel by Lord Lytton called *The Last Days of Pompeii*. However, it was certainly in use during the eighteenth century, when it meant someone who prepared and dispensed medicines, although by the beginning of the nineteenth century most people working in this area would have called themselves chemists or druggists. The profession was unregulated until the formation of the Pharmaceutical Society of Great Britain in 1841.
25. Holloway, 1966, p. 108.
26. Wells, 1799. 'A Letter to Lord Kenyon', footnote to pp. 83–5, referenced in S.W.F. Holloway, *The Apothecaries' Act, 1815: a reinterpretation*.

Chapter 4. The radical Mr Parkinson

1. John Thelwall (1764–1834) and Daniel Eaton (1753–1814).
2. Both James Parkinson and Henry Cline were character witnesses for Thelwall at his trial in 1794. Both said they had then known him seven years, Parkinson adding that for two of these seven years he had known Thelwall 'intimately'. Newton, 1795, p. 53.
3. Minutes of the Physical Society of Guy's Hospital 1775–1825, Kings College London, Special Collections. G/S4/M2. The full details of the child with two heads can be found in Home, 1790.
4. Roe, 2004.

5. Thelwall, 1793.
6. Bynum, 2004.
7. James Parkinson was initiated into the Freemason's Lodge of Freedom and Ease, later known as the Moira Lodge, No. 92, on 15 June 1791, paying a fee of two guineas. He attended meetings quite regularly throughout the remainder of 1791, but then only four times after he joined the LCS in January 1792, the last time being 27 March 1793.
8. Harland-Jacobs, 2010.
9. Parkinson may have been a member of the Friends to the Liberty of the Press at this time for his name is among more than 500 people who signed the Friends' Declaration. *Proceedings of the Friends to the Liberty of the Press, for 22 December 1792, 19 January 1793 and 9 March 1793*. Parkinson's name can be found on page 12 of the Declaration, but not all copies of the *Proceedings* include the Declaration.
10. Thale, 1983. p. vii.
11. Morris, 1989, p. 26.
12. London Corresponding Society, 1792.
13. Daniel Isaac Eaton (1753–1814)
14. Davis, 1995, p. 8.
15. Eaton's new shop was at 81 Bishopsgate Street Without, in the parish of St Boltolph. By the late eighteenth century, stationers and booksellers lined the entire length of Bishopsgate Street. (Davis, 1995, p. 79.)
16. McCue, 2004, p 48.
17. Thale, 1983, p. 83
18. Hay, 2004.
19. Eaton, 1793a.
20. Burke's *Reflections on the Revolution in France* is 356 pages long, but it is largely remembered for this short sentence on page 117.
21. Thomas Spence (1750–1814)
22. James Parkinson writing as Old Hubert in Issue 1 of *Hog's Wash; or, a Salmagundy for Swine*, p. 11.
23. The eighth issue of *Politics for the People* was published on 16 November 1793.
24. John Gurney (1768–1845)
25. Barrell and Mee, 2006–7.
26. Thale, 1983, p. 88.
27. The *Surprize* left England on 2 May 1794, bound for Botany Bay, Australia, with 33 male and 58 female convicts on board.
28. This meeting raised £18 5s for the families of the men about to be transported to Australia. Thale, 1983, p. 185.
29. The *Habeas Corpus* Act was passed by English Parliament in 1679 only due to a miscounting of votes. It guaranteed that a person detained by the authorities would have to be brought before a court of law within a certain time frame

and a reason produced for their detention. However, in times of social unrest, Parliament has the power to suspend *Habeas Corpus* and detain suspects as long as deemed necessary. In this instance, the period of suspension lasted from May 1794 to July 1795.

30. Eaton, 1812, p. 1.

Chapter 5. The Pop Gun Plot

1. Parkinson, 1805.
2. Lovat-Fraser, 1932, p. 52.
3. Anon. (Parkinson), 1794a. The fund was first advertised on 19 June 1794.
4. Barrell, John. 1998, p 19.
5. The printing expenses of *Revolutions Without Bloodshed* had been two guineas, so there remained a balance of £1 14s to be donated to the fund for wives and children. Thale, 1983, p. 199. Details of various LCS meetings are also taken from this work.
6. By November 1794, six months after the arrests, subscriptions to the fund totalled £314 19s 3d, of which £284 7s 11½d had been expended, mostly in donations to the families of the men in prison. The fund also provided some legal assistance for the prisoners, but as raising money for that was illegal, it was not mentioned in the accounts (Thale, 1983, p. 224).
7. Lemaitre, 1795.
8. Smith, 1795. Details of Smith's and Higgins's arrest and trial, as well as Parkinson's correspondence with Smith, are taken from this work.
9. *Whitehall Evening Post*, Saturday 11 October 1794.
10. Smith, 1795, p. 45.
11. Wilkes, 2016.
12. Gurney, 1796.

Chapter 6. Trials and other tribulations

1. This meeting was held on 3 October 1794, in Academy Court, Chancery Lane.
2. Thale, 1983, p. 227
3. John Scott, 1st Earl of Eldon (1751–1838) was Attorney General between 1793 and 1799, thus he presided over all the major treason trials.
4. When Smith was asked the reverse question – how long he had known Parkinson – he answered ten or twelve years. It is possible the court stenographer misheard one or other of their replies.
5. All the details of this trial can be found in Smith, 1795.
6. Smith, 1795, p. 69
7. This was the trial in Edinburgh of Robert Watt and David Downie in August 1794. Watt was executed and Downie exiled.
8. Eaton, 1793b. p. 19.
9. Advert in *The Times*, 15 December 1794, p. 2.

10. Fauvell and Simpson.
11. Smith, 1795, p. 65.
12. Lemaitre, 1795.
13. Upton was actually the first to be released on bail, on 6 May 1795. National Archives, Kew. Records of the Court of King's Bench and other courts: KB 33/6/4
14. Old Hubert (Parkinson), 1795a.
15. Old Hubert (Parkinson), 1795b.
16. One cannot help being reminded of a certain MP who claimed expenses of £1,645 for a floating duck house on his pond.
17. Paul Lemaitre relates this information in his second petition, dated 14 August 1846; however, there is no evidence in the transcript of the trial that suggests these witnesses came forward. It may be that, 50 years later, Lemaitre was misremembering events. This petition, and the first one, can be found at the end of *Some remarks respecting the supposed origin of the Pop Gun Plot by P.T. Lemaitre, 27 September 1833.* British Library: Place Papers, British Museum Add MS 27808.
18. Emsley, 1979, p. 552.
19. Smith, 1795, pp 41-77.
20. London Corresponding Society, 1795, p. 8.
21. Thale, 1983, p. 322.
22. Anon., 1797, p. 35.
23. *The Telegraph*, 1 July 1796, p. 3.
24. Lemaitre. Second petition. British Library: Place Papers, British Museum Add MS 27808.
25. Stuart Jones, 1950.
26. Davis, 2004.
27. Hardy, 1832, p. vii.

Chapter 7. Dangerous sports

1. Parkinson, 1800b, p. 70.
2. The word rabies comes from the Latin meaning 'madness'. It is a virus that affects the brain and spinal cord and is usually transmitted through the saliva of an infected animal into the wound caused when bitten. The virus takes two to eight weeks to incubate before signs of the disease are noticed. Early symptoms are headache and fever, followed by uncontrollable spasms and hydrophobia, eventually leading to coma and death. In Parkinson's time, once symptoms appeared, death was inevitable.
3. Parkinson, 1814a.
4. Parkinson says he had given his notes on this case to a Dr Marshall at the London Hospital, in the hope that Marshall would publish them, but this had not happened by the time Parkinson published his work on hydrophobia.

They were eventually included in Dr Marshall's *The Morbid Anatomy of the Brain, in mania and hydrophobia* which was posthumously published in 1815. Marshall, however, makes no reference at all to Parkinson's suggestions as to how the woman might have contracted the disease.

5. Pearn and Gardner Thorpe, 2001.
6. Parkinson, 1807a.
7. Dodd, 1841.
8. 'Report of the Select Committee on Factory Children's Labour', 1833. This was usually referred to at the time as 'the report of Mr Sadler's Committee' because it was written by Michael Sadler, the Chairman of the Parliamentary Committee.
9. Parkinson, 1807a.
10. The full list of regulations can be found in Morris 1989, pp. 83–4.
11. Parkinson, 1800e.
12. Sterne, 2008, p. viii.
13. Mackenzie, 1967.
14. Parkinson, 1800b, p. 85
15. Parkinson, 1799, p. 485.
16. Parkinson, 1799, p. 389.
17. Clayton, 1807.

Chapter 8. A pox in all your houses

1. Dunn, 2000.
2. Buchan, 1790, p. 54
3. Parkinson, 1799. Admonitions meant cautionary advice, rather than a rebuke.
4. The word malaria is derived from the Italian *mala aria* which means 'bad air'.
5. Reiter, 2000.
6. Parkinson, 1800b, p. 83.
7. Anon., 1804, pp. 265–8.
8. Parkinson, 1802.
9. Parkinson, 1799, p. 234–9.
10. Jenner, 1798.
11. Parkinson, 1799, p. 499
12. Minutes of the Board of Directors of the Royal Jennerian Society for the Extermination of the Small Pox. 1802–1805. Entry for Friday 3 December 1802. Wellcome Library, MS.4302.
13. Minutes of the Board of Directors of the Royal Jennerian Society for the Extermination of the Small Pox.1803–1806. Entry for 17 February 1803. Wellcome Library, MS.4303.
14. Medical Committee minutes. 7 September 1803 and 7 December 1803. Wellcome Library, MS.4304.

15. Parkinson was co-opted on to the Royal Jennerian Society Medical Committee on 7 March 1804.
16. General Court minutes. Thursday 25 February 1808. Wellcome Library, MS.4306.
17. *Trewman's Exeter Flying Post or Plymouth and Cornish Advertiser* (Exeter, England), Thursday 12 November 1807; Issue 2301. The same article appeared in many other papers.
18. Lewis, 2009a, p. 67

Chapter 9. The fossil question

1. *General Evening Post* (London, England), 29 May 1788–31 May 1788; Issue 8508.
2. *St James's Chronicle or the British Evening Post* (London, England), 26 June 1788–28 June 1788; Issue 4233. *Morning Chronicle and London Advertiser* (London, England), Tuesday 24 June 1788; Issue 5967.
3. *General Evening Post* (London, England), 29 May 1788–31 May 1788; Issue 8508.
4. Display notice in the Hunterian Museum, Royal College of Surgeons.
5. Moore, 2005, p. 469.
6. Parkinson's response to the award of the Royal College of Surgeons' Gold Medal. Minutes of the Council of the Royal College of Surgeons, Friday 11 April 1823. Court of Assistants and Council Minute Book. RC S-GOV-1-1-3.
7. Rappaport, 1997, p. 106.
8. James Hutton (1726–1797).
9. John Hunter wrote *Observation and Reflections on Geology* around 1790 although it was not printed until 1859, more than sixty years after he died. In the preface, he writes: 'As the Fossils of the sea, or water-animals, can now only be found upon land, it is a proof that the sea was once there, and from this alone we may presume that where the sea now is, it was once land. This leads to two modes of the exposition of the Earth: one, the sea leaving the land; and the other, the bottom of the sea rising up above the water by some convulsive motion of the earth . . .' (p. xlvi).
10. Parkinson, 1811c, p. 432.
11. Porter, 1982, p. 184.
12. De Luc, 1779.
13. Rudwick, 1997, p. 24.
14. Ibid., p. 17.
15. Lamarck, 1809.
16. Stott, 2012, p. 300.
17. Rudwick, 1997, p. 21.
18. Parkinson, 1808, p. 110.

19. *Alcyonium* is a genus of soft corals. Those referred to here can be seen in Parkinson, 1808, Plate X, Figs 14–16.
20. A number of the catalogues from the auction sales Parkinson attended between 1798 and 1808 have survived, giving us an insight into his range of interests and his willingness to spend more and more on the collection. These are held in the library at the Natural History Museum.
21. Parkinson, 1804, Preface.
22. Parkinson, 1804, p. 30.
23. The Royal College of Surgeons still holds the letter Parkinson wrote in 1800 'earnestly entreating permission' to be allowed access to Hunter's collection and 'derive that aid to his pursuits which so grand and interesting a collection may furnish'. Archives of the Royal College of Surgeons: RCS-MUS/5/6/1, 22 Nov 1800.
24. Klein, 2004.
25. Parkinson, 1800a, p. 52.
26. Parkinson, 1800d.
27. See, for example, the advert in *Morning Chronicle*, 24 October 1807.
28. The notice faces page 251 in the second edition of Parkinson's *Chemical Pocket-Book*, published in 1801.
29. Calomel was mercury (I) chloride, having the formula Hg_2Cl_2. It was often given as a laxative but when combined with opium, that ameliorated the laxative effects. The World Health Organization has deemed mercury unsafe at any level of exposure.
30. Although not officially a mainstream medication for gout, one only has to look on the internet to see how many people with the disease find taking bicarbonate of soda beneficial.

Chapter 10. A sublime and difficult science

1. Welch was the son of a wealthy business man, another Wakelin Welch, who died at Maryland Point in the County of Essex, in Virginia, America, and who was described as a 'merchant'. From at least 1780, the business of Wakelin & Son is recorded in trade directories as being located at 10 Fenchurch Buildings, London and it must have been during this time that Parkinson got to know his friend. Of interest is the fact that the father corresponded with George Washington regarding goods Washington required sending to America. These included a number of agricultural implements because, as Washington explains, 'agriculture has ever been among the most favourite amusements of my life, though I never possess much skill in the art; and nine years total inattention to it has added nothing to a knowledge which is best understood from practice'. On another occasion Washington asked Welch to arrange the sending of 'Twenty six dozen [bottles] of claret and the same quantity of champagne'. When the father died in 1796, Welch continued the family business, eventually

leaving a large legacy that was used, among other things, to fund the Powell and Welch Almshouse Charity, which still exists today. The almshouses in Bitteswell, Leicestershire, built in 1847, have the inscription: 'These almshouses were built from funds left for charitable purposes by the late Wakelin Welch, Esq. of Camden Place, Bath, and Elizabeth his wife, sister of the late Rev James Powell, vicar of this parish.' Welch died in 1818 aged 58.

2. Welch, 1822. The book was not published until after Welch's death in 1818, so it may well have been written some considerable time earlier, which might account for its outmoded ideas.
3. There is some confusion as to Wilton's name, since later on in the volume he is called 'Winton'. I have chosen to use the first version, assuming the latter to be a typographical error.
4. Parkinson, 1804, p. 3.
5. Parkinson, 1811c, p. 449.
6. Letter from James Parkinson to Jean-André de Luc, 11 December 1812. University of Bristol Library Special Collections. Bristol University library and several others hold photographs of the original letter – if anyone knows the whereabouts of the original, please get in touch.
7. O'Connor, 2007, p. 361.
8. Parkinson, 1811c, p. 449.
9. Hunter, 1859. Hunter's remark p. iii, and Renell's comment p. lvii. Hunter duly made the recommended alteration. See: Quist, George. 1979.
10. Priestley, 1768, p. 16.
11. Parkinson, 1811c, p. 464.
12. Parkinson, 1804, p. 469.
13. Parkinson, 1811c, p. xiv.
14. Parkinson, 1804, p. 471.
15. *The British Critic.* 1805. Vol. XXVI, pp. 1–17.
16. *The Eclectic Review.* 1805. Vol. I, pp. 44–7. It is interesting to note that this reviewer uses the word mineralogist when he means geologist, reinforcing the fact that the word geology was still not in common use at this time.
17. Arthur Aikin (1773–1854). This is the correct spelling of his name, although various online sources give Aiken.
18. *The Annual Review, and History of Literature.* 1805. Vol. III, pp. 904–906. As early as 1796 the editor, Arthur Aikin, had undertaken a geological tour of Wales and Shropshire. Like Parkinson, he was to become one of the thirteen founders of the Geological Society.
19. Wollaston, 1797.
20. Pearson, 1798.
21. In these modern times gout can also be caused by medications that interfere with the body's ability to remove uric acid, such as aspirin, diuretics and levodopa, which, ironically, is used to treat Parkinson's disease. Also of interest is

the fact that people who have gout are less likely to develop Parkinson's disease because uric acid exerts antioxidant effects on neurons (De Vera, *et al.* 2008.)

22. Letter from Greenough to Pepys, 16 June 1807. B20, Pepys Collection, Royal Institution.
23. Parkes, 1815, p. 125.
24. Geological Society of London, Greenough Papers: LDGSL 960.
25. In the event, eleven men sat down to dine at five o'clock on 13 November 1807: Arthur Aikin (1773–1854), William Allen (1770–1843), William Babington (1756–1833), Jacques Louis Comte de Bournon (1751–1825), Humphry Davy, James Franck (1768–1843), George Bellas Greenough (1778–1855), Richard Knight (1768–1844), James Laird (1779–1841), James Parkinson (1755–1824) and Richard Phillips (1778–1851). William Hasledine Pepys (1775–1856) and William Phillips (1773–1828) were unable to attend but, due to having been invited, they have always been considered founder members. For a full account of the founding of the Geological Society, see Lewis, 2009a.
26. Allen, 1847, p. 66.
27. *Geological Inquiries.* GSL: LDGSL 352. A copy of this pamphlet is now available as Appendix I in Lewis & Knell, 2009.
28. Geological Society of London: Ordinary Meeting Minutes 1807–1818. Members of Committee of Maps were: William Atkinson (1774/5–1839), Rev. Edward John Burrow (1785–1861), [Gilbert] Giddy Davies (1767–1839), James Franck, Leonard Horner (1785–1864), Wilson Lowry (1760–1824), John MacCulloch (1773–1835), William MacMichael (1783–1839), William Pepys, William Phillips, James Parkinson, and James Shuter (dates unknown).
29. One of Smith's maps recently sold for £100,000; copies of Greenough's map go for much less.
30. Simon Winchester's book *The Map That Changed the World* (London: Viking, 2001) is a biography of William Smith.
31. Parkinson, 1808, p. 19.
32. *The Annual Review, and History of Literature for 1808.* 1809. Vol. VII, p. 709.
33. These experiments are referred to in Parkinson, 1804, letter XXXVII, p. 353, and in an undated letter (probably May 1804) from James Parkinson to William Hasledine Pepys, held at the Royal Institution, London: Pep/B/49.
34. Lewis, 2000.
35. Grateful thanks are due to Chris Toland for suggesting that elaterite could be the substance Parkinson had revealed.
36. Mantell, 1850, p. 87. Thanks to Gregory Todd for pointing out Mantell's comments on this experiment.
37. Miller, 1821. Parkinson was a subscriber to this volume, in fact the list of subscribers reads like a who's who in geology of the time: The Geological Society, Henry de la Beche (1796–1855), Jacques Louis Comte de Bournon, Richard

Bright (of Bright's disease: 1789–1858), Rev. William Buckland (1784–1856), Rev. William Conybeare (1787–1857), George Greenough, Gideon Mantell, William Phillips, Adam Sedgwick (1785–1873), George Brettingham Sowerby (1788–1854) and many other eminent geologists.

38. Thackray, 1976.
39. Thackray, 1976, p. 452.

Chapter 11. 'Tis a mad, mad world in Hoxton

1. *Courier & Evening Gazette*, 14 October 1794.
2. Norris Brewer, 1816.
3. The name of the owner of Holly House is variously recorded as Burrow, Burrows, Burrowes and Burroughs.
4. House of Commons, 1816, p. 128.
5. *The Universal Magazine*, New Series, 1811. Vol. XV. p. 133.
6. Ibid., p. 131.
7. John Mitford (1732–1831) was confined in Whitmore (Warburton's) House from May 1812 to March 1813.
8. Mitford, 1825, p. 3.
9. Rogers, 1815.
10. Parkinson, 1811a.
11. All quotations in this paragraph are from the *Hull Packet*, 13 November 1810.
12. The date of articles in *The Morning Chronicle* and *The Times* is 30 October 1810.
13. *The Statesman*, 31 October 1810. Editorial on the Daintree case.
14. *The Statesman*, 14 November 1810. Letter to the Editor.
15. Parkinson, 1811a.
16. Rogers, 1815.
17. House of Commons, 1816.
18. *The Universal Magazine*, New Series, 1811. Vol. XV. pp. 133–6; *The Gentleman's Magazine*, New Series, 1811. Vol. LXXXI. pp. 254–5; and *The Monthly Review*, 1811. Vol. LXVI. p. 103.
19. *The Monthly Review*, 1811.
20. Geological Society of London: Ordinary Meeting Minutes 1807–1818. Members of the Committee for Nomenclature were: Arthur Aikin, Humphry Davy, George Bellas Greenough, Sir James Hall (1761–1832), Leonard Horner, James Laird, Wilson Lowry, William MacMichael, John McCulloch and William Phillips.
21. For an example of how two completely different rocks called by the same name led David Mushet to think there was another coalfield beneath the one in the Forest of Dean, see Lewis, 2016.
22. Parkinson also served on the Society's Council between 1813 and 1815.
23. Peter Roget (1779–1869).

24. University College London archives: AD7981 831. Horner to Greenough, 7 August 1811.
25. Parkinson, 1811b.
26. Quoted in Morris, 1989, p. 124.
27. Parkinson, 1811c.
28. Farey, 1811. For more on Farey's significant contribution to the early understanding of British geology see Lewis, 2016, and references therein. A more recent estimate of the amount of overburden removed from northern England is 3km, rather than 3 miles, but Farey was certainly on the right track. See Lewis *et al*, 1992.
29. Challinor, 1948.
30. *The Monthly Review*, 1813. Vol. LXX. p. 20.
31. *The Monthly Magazine*, 1811. Vol. XXXII, part II. p. 694.
32. Heringman, 2004, pp. 73, 162.
33. Higgins, 2011.
34. Lewis, 2013.
35. Lewis, 2009b.

Chapter 12. The name of the father, and of the son

1. Parkinson, 1805, p. iv.
2. John Parkinson, 1811.
3. An early eighteenth-century recipe for turpentine glyster was as follows: Take Urine of a Man in Health 1 pint; Venice Turpentine (dissolved in 2 Yolks of Eggs) 1 ounce; Oil of Aniseed 1 dram; Melassos 1 ounce, mix. (Thomas Fuller, *Pharmacopoeia extemporanea: or, a body of prescripts. In which forms of select remedies, accommodated to most intentions of cure, are propos'd* [B. Walford, London: 1710].)
4. John Parkinson, 1812.
5. There is a report from 1735 in the *Transactions of the Royal Society* describing a complex operation performed by Claudius Amaynd, Sergeant Surgeon to his Majesty, on a young boy who had swallowed a pin, which required the removal of his appendix. Notwithstanding that the operation proved 'the most complicated and perplexing I ever met with', and lasted half an hour (without anaesthetic), which the young patient bore with 'great courage', it was a complete success and the boy fully recovered. So while John Parkinson's account may have been the first to record the pathology of a burst appendix, Amaynd was the first to record removal of an appendix.
6. The duties paid per cwt at that time were: '£1 12s. 8d. for plate glass, 8s. 2d. on what is called Streadwinter's glass, and 4s. 8d. on crown glass'. Supplementary Budget. House of Commons debate 22 March 1805, Hansard vol. 4, pp. 87–90.
7. Suzuki, 2004.

8. *Transactions of the associated apothecaries and surgeon-apothecaries of England and Wales*, 1823. Vol. 1.
9. 'Review of Observations on the Necessity of Parochial Fever Wards', *The London Medical and Physical Journal*, 1819. Vol. 41, Issue 241. p. 270.
10. A copy of this letter can be found in Morris, 1989, p. 84.
11. For a fuller version of these reviews, see Morris, 1989, p 85.
12. 'Medical and philosophical intelligence', *The London Medical and Physical Journal*, 1819. XLI (3).
13. *Morning Chronicle*, 12 September 1818.
14. William Maiden had studied under Astley Cooper at St Thomas's Hospital and had been with him in revolutionary Paris where they witnessed the atrocities of the Terror.
15. Anon., 1813.
16. Maiden, 1812, p. 25.
17. Wood, 1960.
18. Cooke, 1835, p. 9.

Chapter 13. The shaking palsy

1. Parkinson, 1817.
2. Gardner Thorpe, 1987, contains a facsimile of Parkinson's *Essay*.
3. Parkinson's use of the term 'species' here when referring to a disease is interesting. Thomas Sydenham (1624–1689), known as the 'English Hippocrates', had, in the previous century, urged medical practitioners to identify and classify diseases with the same rigour as botanists characterised plants. Sydenham's idea was to take diseases as they presented themselves in nature and to draw up a complete picture of the specific characteristics of each, which is exactly what Parkinson was trying to do with his *Essay*. Most forms of ill-health, Sydenham argued, had a definite type, comparable to the types of animal and vegetable species. From Parkinson's knowledge of plants as an apothecary, he would have been familiar with classifying plants in this way – fossils too, of course – and it would have seemed natural to assemble all the symptoms of the shaking palsy as defining a single 'species' of disease.
4. The earliest reference to publication of Parkinson's *Essay* appeared in an advert placed in the *Morning Chronicle* of 31 May 1817, under a list of books 'published this day'; price 3s. On 5 June the same advert appeared again, also announcing the *Essay* had been 'published this day'.
5. Morris, 1989, p. 138.
6. Elliotson, 1831, p. 119.
7. Watson, 1855, p. 411.
8. Maclachlan, 1863, p. 217.
9. Sanders, 1865.
10. Buzzard, 1882.

11. Jean-Martin Charcot (1825–1893).
12. Goetz, 1986.
13. Charcot, 1888, p. 175E.
14. Goetz, 2011.
15. Charcot, 1872, p. 133.
16. The substantia nigra was discovered by a French physician, Félix Vicq d'Azyr (1748–1794), in 1784.

Chapter 14. Reforms and rewards

1. *Chester Courant*, 10 December 1816.
2. *Habeas Corpus* was suspended between January 1817 and March 1818.
3. Morris, 1989, pp. 77–9.
4. Letter to William Buckland from James Parkinson, 28 January 1821. Oxford University Museum of Natural History. Buckland Letters, Box 2/P3.
5. Parkinson, 1822.
6. The notice of its withdrawal can be found in: Nicholson, William, *Journal of Natural Philosophy, Chemistry, and the Arts*, 1810. Vol. 25, p. 317. London.
7. Royal College of Surgeons, Board of Curators Minutes Book, RCS-MUS213 Vol. 3, 3 October 1822.
8. Parkinson's 68th birthday was on 11 April 1823.
9. Royal College of Surgeons, Court of Assistants and Council Minute Book, RCS-GOV-1-1-3, 11 April 1823.
10. Anon., 1824.
11. Parkinson and Parkinson, 1824.
12. Minutes of the Board of Trustees of the poor of the parish of St Leonard's, 23 December 1824. Cited in Morris, 1989, p. 12.
13. At the time of his death, Parkinson had property, 'messuages and tenements', in Stratford Langthorne, Essex; perhaps they had been bought with the proceeds from his various publications. The rent from these was to be shared between his wife and his two sons, John and Henry. His youngest daughter Mary would inherit his wife's share when she died, although this was to be forfeited if she married, which she did four years later. She did however get £500 'for her own use and benefit', as did her sister Emma who was already married. Parkinson's brother-in-law and good friend, John Keys, was given 'my cabinet of coins and medals' and Parkinson's sister Mary, John Keys's wife, received a diamond brooch to be worn 'in memory of our affectionate brother'. This must have been their brother William who had died in 1766 aged five. Parkinson's second son, Henry, had gone to live in Calcutta and James had evidently paid for Henry's fiancée, Elizabeth Asperne, to join him so they could marry there in 1820. Henry was expected to return the cost of the fare, £134 13s 6d, but James bequeathed this to Elizabeth, rounding it up to £138. (Henry and Elizabeth

did return to England at some point, where they died, but they appear not to have had any children.)

14. Royal College of Surgeon's Board of Curators Minutes Book, Vol. 3: various dates after 4 December 1825, RCS-MUS213.
15. George Brettingham Sowerby I (1788–1854).
16. *Morning Chronicle*, 2 April 1827.
17. Mantell, 1846, p. 14.
18. Sedgwick Museum Archive, Cambridge, DDF859. Copies of the Annual Reports of the Woodwardian Inspectors 1779–1862.
19. Personal communication from Matt Riley, Sedgwick Museum Archive, Cambridge. Email dated 3 September 2013.
20. Those purchased by James De Carle Sowerby (1787–1871), Matthew Wright of Bristol, Gideon Mantell, and William Willoughby Cole, 3rd Earl of Enniskillen, were later transferred to the British Museum of Natural History, although some of Enniskillen's went to the Geological Survey Museum. E.W. Swanton noted that the collection at the Haslemere Museum in Hampshire 'contains some of the specimens used by Parkinson in illustrating his "Organic Remains"'. Cleevely, 1983, p. 224.
21. Mantell, 1846, p. 14.
22. Mantell, 1850, p. 126.
23. Parkinson, 1811c, p. 38.
24. Mantell, 1850, p. 115.
25. Allingham, 1924, p. 45.
26. This sponge was originally named after Parkinson by Michelin, subsequently renamed *Spongites townsendi* by Mantell, and is now called *Chenendopora michelinii* Hinde (NHM P.1197). It was originally labelled, presumably by Parkinson, as being from the Cretaceous, Upper Greensand, of Wiltshire, England, but it is now considered to be from France. Personal communication: discussion with Consuelo Sendino, Curator, Earth Sciences Department, Natural History Museum.
27. If anyone knows the whereabouts of a catalogue from the sale of Parkinson's collection, please let me know.
28. Letter from John Parkinson to George Sowerby, 4 March 1828. George Brettingham Sowerby I Letters, National Museum of Wales, Cardiff.
29. When John Parkinson died he was living at 11 Suffolk Place, Lower Road, Islington East.
30. Mary Parkinson died on 28 March 1838 at 31 Aberdeen Place, Marylebone. She was 80.
31. Kohn, 2007, p. 107.
32. Watkins & Schoberl, 1816, p. 262.
33. These quotes are from various reviews that were compiled into a whole page advert for many of Parkinson's books. It can be found at the back of the 1820

edition of volume 2 of *Organic Remains*. It is interesting to note his books were still selling twenty years after some of them had been written.

34. The horticulturalist was J.W.S. Van der Wereld.
35. Mantell, 1850, p. 13.
36. Five hundred copies had been printed of Parkinson's first, 1804, volume of *Organic Remains*. Printing costs, including 1,500 copies of advertising prospectuses, came to £118 11s 6d. This first volume sold for £2 2s 6d, rising to £2 12s 6d in 1808, when the second volume was published, which also sold for £2 12s 6d. The third volume, published in 1811, was more expensive at £3 13s 6d. All three volumes could then be purchased for £8 18s 6d, saving the purchaser a pound on buying the volumes separately. Various editions of each volume became available over the next 30 years until around 1844 a 'second edition' became available for ten guineas, with 'extra cloth boards £4:4:0'. Full details of all these editions can be found in Thackray, 1976.
37. Frank Buckland, for example, stated how his father William Buckland had assimilated 'everything that illustrated the new science of fossil organic remains, then just coming into vogue through the work of Parkinson'. Buckland, F., 1858, p. xxv.
38. Mantell, 1850, pp. 119 and 31 respectively.
39. James Sowerby (1757–1822). The Sowerbys were a family of several generations of naturalists and illustrators who were active from the late 18th century. Of those mentioned here, James was the father of the auctioneer George Brettingham (see page 235), and James De Carle (see note 20, above), who purchased some of Parkinson's fossils.
40. Mantell, 1850, p. 125. Mantell further mentions a number of other fossils that bear Parkinson's name, including the *Nautilus parkinsoni*, named by 'Mr Edwards' in recognition of Parkinson's research on the mechanisms of the nautilus's siphuncle (now called *Aturoidea parkinsoni*); a fossil starfish called by Parkinson *Goniaster semilunata*, but renamed *Goniaster parkinsoni* by Edward Forbes in 1848, and 'A beautiful example of the Turban Echinite (*Cidaris parkinsoni*, of Dr. Fleming), from Wiltshire'.
41. The Honorary Medal is still awarded at irregular intervals to candidates who meet exactly the same criteria as when Parkinson received it all those years ago – 'liberal acts or distinguished labours' – but in the 35 years between 1834 and 1869 no awards were made at all, the College adamantly adhering to its policy of granting it only to contenders of outstanding merit.
42. Parkinson, 1822, p. 334.

Epilogue: A fragment of DNA

1. James Warren Parkinson (1837–1900).
2. *The London Medical and Surgical Journal*, Vol. V, 1834, p. 608.
3. Sarah Anne Warren married again in 1866.

4. Quite how this commission was purchased is unclear, for they cost many hundreds of pounds. Perhaps Sarah Anne's new husband paid for it; as a lieutenant in the Navy, he would have had the appropriate connections and funds.
5. Although only 736 tons (register), the *May Queen* made passages to several ports. She ran to Dunedin from 1871 until 1876, under charter to the Shaw Saville Company. After making sixteen voyages to New Zealand she came to grief at Lyttelton in 1888, while in the command of a Captain Colville, who had made six voyages in the ship. http://freepages.genealogy.rootsweb.ancestry.com/~ourstuff/MayQueen1872.htm
6. Issue of *The May Queen Weekly News* published on board the vessel en route from London to Otago (16 September 1874). fMS-Papers-7615, National Library of New Zealand.
7. *Otago Daily Times*, Issue 3977, 14 November 1874. Page 2: Arrival of the *May Queen*.
8. W.D. Borrie, 1991, p. 52.
9. John Reynolds, his father Robert, and other Parkinson descendants have created an excellent genealogy website, which I found invaluable: http://reynoldsfamilynz.tribalpages.com
10. The Richter scale is a logarithmic scale, so each whole unit is ten times larger than the one before it.
11. Gideon Mantell's son, Walter, also immigrated to New Zealand, in 1839, from where he sent back some important fossils.
12. When we shake hands we exchange cells, some of which contain our DNA, and since every child receives half its DNA from each parent, John Reynolds's DNA must contain some inherited from Parkinson. Fragments of DNA can be sequenced from about two-thirds of samples found on an inanimate object after someone has touched it. They are thought to derive from secretions from sweat and sebaceous glands, as well as being associated with cells and fragments of cells. The chances of transfer are increased by friction between the hand and surfaces, and so would be higher after shaking hands, compared to a simple touch.

BIBLIOGRAPHY

Works by the Parkinsons (ordered chronologically)

Parkinson, John (Snr), 1774. *Transactions of the Royal Humane Society; dedicated by permission to His Majesty by W. Hawes*, Vol. 1, pp. 161–2.

Parkinson, James, 1789. 'Some account of the effects of lightning' in *Memoirs of the Medical Society of London*, Vol. 2, pp. 493–507.

Z.A. (Parkinson, James), 1792. Letter to *The Gentleman's Magazine*. Vol. LXII (April), pp. 291–2.

Parkinson, James, 1792. Letter to *The Gentleman's Magazine*. Vol. LXII (October), pp. 897–9.

Old Hubert (Parkinson, James), 1792–3. *The Budget of the People.* Parts 1 and 2. Printed for Daniel Isaac Eaton, No. 81 Bishopsgate without, London.

Old Hubert (Parkinson, James), 1793. *An Address to the Hon. Edmund Burke from the Swinish Multitude.* Printed for J. Ridgway, York Street, St James's Square, London.

Old Hubert (Parkinson, James), 1793. *Pearls Cast Before Swine by Edmund Burke, scraped together by Old Hubert*. Printed for Daniel Isaac Eaton, No. 81 Bishopsgate Without, London.

Old Hubert (Parkinson, James), 1793. *Knave's-Acre Association. Resolutions adopted at a meeting of placemen, pensioners etc, held at the Sign of the Crown, Knave's-Acre.* Printed for T. Spence, No. 8, Little Turn-Stile, High-Holborn, London.

Old Hubert (Parkinson, James), 1793. *The Village Association, or the politics of Edley.* Printed for J. Ridgway, York Street, St James's Square, London.

Old Hubert (Parkinson, James), 1793. *The Soldier's Tale. Extracted from the Village Association with two or three words of advice by Old Hubert.* (2nd edition). Printed by Daniel Isaac Eaton, No. 81 Bishopsgate Without, London.

Old Hubert (Parkinson, James), 1793–5. Various articles in: *Politics for the People, or a salmagundi for swine*, 2 vols. Daniel Eaton, London.

Old Hubert (Parkinson, James), c. 1794. *Mast and Acorns: collected by Old Hubert.* Printed for Daniel Isaac Eaton, 74 Newgate-Street, London.

Anon. (Parkinson, James), 1794a. *Revolutions Without Bloodshed, or reformation preferable to revolt.* Daniel Eaton, Newgate-Street, and John Smith, Portsmouth Street, Lincoln Inn Fields, London. (Parkinson admitted he had written this during his interrogation by the Privy Council.)

Anon. (Parkinson, James), 1794b. *A Vindication of the London Corresponding Society.* Sold by J. Smith, Portsmouth Street, Lincoln Inn Fields, and J. Burks, No. 52 Crispin Street, Spitalfields, London. (Parkinson also admitted he had written this during his interrogation by the Privy Council.)

Anon. (Parkinson, James), 1794c. *Reformers No Rioters, containing remarks on the conduct of magistracy towards groups and kidnappers.* Printed by order of the London Corresponding Society, London.

Anon. (Parkinson, James), 1795a. *Ask, and You Shall Have, or the source of public grievances displayed and their remedies pointed out. An address to the people.* Printed and sold by J. Burks, Crispin-Street, Spitalfields, London. (Listed by Smith as being by Old Hubert.)

Old Hubert (Parkinson, James), 1795b. *A Sketch. While the Honest Poor Are Wanting Bread.* J. Burks, Spitalfields, London.

Parkinson, James, 1795c. Letter to Mr Smith, bookseller, Portsmouth-Street, Lincoln's Inn fields. In: *Assassination of the King! The conspirators exposed, or, an account of the apprehension, treatment in prison, and repeated examinations before the Privy Council, of John Smith and George Higgins, on a charge of High Treason.* J. Smith, at the Pop-Gun, Portsmouth-Street, Lincoln's Inn Fields.

Old Hubert (Parkinson, James), 1795d. *An Account of Some Peculiar Manners and Customs of the People of Bull-Land, or the Island of Contradictions; faithfully detailed by Old Hubert.* Printed for J. Smith at the Pop-Gun, Portsmouth Street, Lincoln's Inn Fields.

Parkinson, James, 1799. *Medical Admonitions addressed to families respecting the practice of domestic medicine, and the preservation of health.* (2 vols.) Printed for C. Dilly, Poultry, London.

Parkinson, James, 1800a. *The Hospital Pupil; or, an essay intended to facilitate the study of medicine and surgery.* Printed for H.D. Symonds, London.

Parkinson, James, 1800b. *The Villager's Friend and Physician; or, a familiar address on the preservation of health and the removal of disease, on its first appearance, supposed to be delivered by a village apothecary. With cursory observations on the treatment of children, on sobriety, industry etc. Intended for the promotion of domestic happiness.* Printed by James Sammells, No. 14, George's Court, Clerkenwell; published by H.D. Symonds, Paternoster-Row, London.

Parkinson, James, 1800c. *The Way to Health* (poster). Printed for H.D. Symonds, No. 20, Paternoster Row, London.

Parkinson, James, 1800d. *The Chemical Pocket-book or memoranda chemica, arranged in a compendium of chemistry, according to the latest discoveries, with Bergman's table of single elective attractions, as improved by Dr. G. Pearson.* Printed by H. Fry, Finsbury Place, and published by D.H. (*sic*) Symonds, Murray, and Highley, Callow, Cox, Arch and Cuthell, London.

Parkinson, James,1800e. *Dangerous Sports, a tale addressed to children.* J.G. Barnard, London.

Parkinson, James, 1802. *Hints for the Improvement of Trusses, intended to render their use less inconvenient and to prevent the necessity of an understrap: with the description of a truss of easy construction and slight expense, for the use of the labouring poor, to whom this little tract is chiefly addressed.* Printed by C. Whittingham, for H.D. Symonds.

Parkinson, James, 1804. *Organic Remains of a Former World. An examination of the mineralised remains of the vegetables and animals of the antediluvian world, generally termed extraneous fossils*, Vol. 1. Printed by C. Whittingham, Dean Street, and published by J. Robson, New Bond Street, London.

Parkinson, James, 1805. *Observations on the Nature and Cure of Gout, on the nodes of the joints, and of the influence of certain articles of diet in gout, rheumatism, and gravel.* London: printed by C. Whittingham, Dean Street; for H.D. Symonds, Paternoster Row; Murray, Fleet Street; Arch, Cornhill; and Cox, St Thomas Street, Borough.

Parkinson, James, 1807. *Remarks on Mr Whitbread's Plan for the Education of the Poor: with observations on Sunday schools and on the state of the apprenticed poor.* H.D. Symonds, London.

Parkinson, James, 1807. *Letter to the Gentleman's Magazine* (77) 2, p. 818.

Parkinson, James, 1808. *Organic Remains of a Former World. An examination of the mineralised remains of the vegetables and animals of the antediluvian world, generally termed extraneous fossils*, Vol. 2. Printed by C. Whittingham, Dean Street, and published by J. Robson, New Bond Street, London.

Parkinson, James, 1811a. *Mad-houses: observations on the Act for regulating mad-houses, and a correction of the statements of the case of Benjamin Elliott, convicted of illegally confining Mary Daintree: with remarks addressed to the friends of insane persons.* Printed by Whittingham and Rowland, for Sherwood, Neeley, and Jones, London.

Parkinson, James, 1811b. 'Observations on some of the strata in the neighbourhood of London, and on the Fossil Remains contained in them', *Transactions of the Geological Society*, 1, 324–54. Printed for the Society by William Phillips, George-yard, Lombard-Street, London.

Parkinson, James, 1811c. *Organic Remains of a Former World. An examination of the mineralised remains of the vegetables and animals of the antediluvian world, generally termed extraneous fossils*, Vol. 3. Printed by C. Whittingham, Dean Street, and published by J. Robson, New Bond Street, London.

Parkinson, John W.K., 1811. 'A case of trismus, successfully treated', *Medical and Chirugical Transactions* (2), pp. 293–7.

Parkinson, John W.K., 1812. 'Case of diseased appendix vermiformis', *Medical and Chirugical Transactions* (3), pp. 57–8.

Parkinson, James, 1814a. 'Cases of hydrophobia', *London Medical, Surgical and Pharmaceutical Repository* (1), pp. 289–92.

Parkinson, James, 1814b. 'Observations on the specimens of Hippurites from Sicily', *Transactions of the Geological Society of London (1st series)*, 2, pp. 277–81.

Parkinson, James, 1817. *An Essay on the Shaking Palsy.* Printed by Whittingham and Rowland for Sherwood, Neeley, and Jones, London.

Parkinson, James, 1818. *Observations on the Necessity for Parochial Fever Wards, with remarks on the present extensive spread of fever*. Sherwood and Co., London.

Parkinson, James, 1821. 'Remarks on the fossils collected by Mr Phillips near Dover and Folkestone', *Transactions of the Geological Society of London (1st series)* (5), pp. 52–9.

Parkinson, James, 1822. *Outlines of Oryctology. An introduction to the study of fossil organic remains; especially of those found in the British strata: intended to aid the student in his enquiries respecting the nature of fossils and their connection with the formation of the Earth.* Printed for the author; and sold by Sherwood, Neely, and Jones, Paternoster Row; and W. Phillips, George Yard, Lombard Street, London.

Parkinson, James and Parkinson, John, 1824. 'On the treatment of infectious or typhoid fever', *London Medical Repository* (1), p. 197.

Parkinson, John W.K. (ed.), 1833. *Hunterian Reminiscences; being the substance of a course of lectures delivered by Mr John Hunter in the year 1785, taken in short-hand and afterwards fairly transcribed, by the late Mr James Parkinson, author of "Organic Remains of a Former World"*. Compton and Ritchie, London.

Other cited works (ordered alphabetically)

Allen, William, 1847. *Life of William Allen: With Selections from His Correspondence* (2 vols). H. Longstreth, Philadelphia, PA.

Allingham, E.G., 1924. *A Romance of the Rostrum: being the business life of Henry Stevens, and the history of thirty-eight King street, together with some account of famous sales held there during the last hundred years*. H.F. & G. Witherby, London.

Anon., 1797. *The Trial of John Smith, Bookseller, of Portsmouth-Street, Lincoln's Inn Fields, Before Lord Kenyon, in the Court of King's Bench, Westminster, on December 6, 1796, for Selling a Work, Entitled, 'A Summary of the Duties of Citizenship!'*. Printed for and sold by Mrs Smith at the Pop Gun, Portsmouth Street, Lincoln's Inn Fields, London.

Anon., 1804. *The Literary Magazine and American Register for 1803–4* (1), pp. 265–8.

Anon., 1813. 'Review: An account of a case of recovery', *The Medical and Physical Journal*, XXIX, 68–71.

Anon., 1814. *Authentic Memoirs, Biographical, Critical, and Literary, of the Most Eminent Physicians and Surgeons of Great Britain: with a choice collection of their prescriptions, an account of the medical charities of the metropolis, &c., &c.* Sherwood, Neely and Jones, London.

Anon., 1818. 'Review of Frankenstein', *Edinburgh Magazine, or Literary Miscellany* (2), pp. 249–53.

Anon., 1824. 'Review: On the treatment of infectious or typhoid fever', *The Medical Recorder* (7), pp. 630–1.

Barrell, John and Mee, Jon (eds), 2006–7. Introduction to: *Trials for Treason and Sedition, 1792–1794*. 8 vols. London: Pickering and Chatto.

Berkowitz, Carin, 2013. 'Systems of display: the making of anatomical knowledge in Enlightenment Britain', *British Journal for the History of Science* (46) 3, pp. 359–87.

Borrie, W.D., 1991. *Immigration to New Zealand, 1854–1938*. Canberra: Demography Program, Research School of Social Sciences, Australian National University.

Bradley, H.J. 1914. *The History of Shoreditch Church*, London. [No publisher details given.]

Buchan, William, 1790. *Domestic Medicine: or, a treatise on the prevention and cure of diseases by regimen and simple medicines*. A. Strahan, London.

Buckland, F., 1858. 'Memoir of William Buckland', in: *Geology and Mineralogy Considered with Reference to Natural Theology*. New edition. Routledge, London.

Burke, Edmund, 1790. *Reflections on the Revolution in France, and on the proceedings in certain societies in London relative to that event. In a letter intended to have been sent to a gentleman in Paris.* London: J. Dodsley in Pall Mall.

Burney, Frances, 1812. Letter to Esther Burney, in *Frances Burney: Journals and Letters*, Sabor, Peter and Troide, Lars E. (eds), 2001. Penguin Classics, London.

Buzzard, Thomas, 1882. *Clinical Lectures on Diseases of the Nervous System*. J. & H. Churchill, London.

Bynum, W.F., 2004. Cooper, Sir Astley Paston, first baronet (1768–1841), *Oxford Dictionary of National Biography*, Oxford University Press, online edn, January 2008.

Challinor, J., 1948. 'The beginnings of scientific palaeontology in Britain', *Annals of Science* (6) 1, pp. 46–53.

Charcot, J.-M., 1872. 'De la paralysie agitante', in: *Oeuvres Complètes de J.M. Charcot (t. 1): Leçons sur les maladies du système nerveux*, pp. 155–88. A. Delahaye, Paris. (English translation: Charcot, J.-M., 1877. 'On Parkinson's disease', in: *Lectures on Diseases of the Nervous System Delivered at the Salpêtrière* [transl. Sigerson G.], New Sydenham Society, London, pp. 129–56).

Charcot, J.-M., 1888. *Oevres Complètes de J.M. Charcot (t. 6): Leçons sur les maladies du foie et des reins*, A Delahaye, Paris.

Cherington, M., 2004. 'James Parkinson: links to Charcot, Lichtenberg and lightning', *Archives of Neurology* (61), p. 977.

Clark-Kennedy, A.E., 1962. *The London: a study in the voluntary hospital system.* Pitman Medical Publishing Co. Ltd., London.

Clayton, 1807. 'On Ching's Worm Lozenges', *The Medical and Physical Journal* (17), p. 173.

Cleevely, R., 1983. *World Palaoentological Collections*, British Museum (Natural History)/Mansell, London.

Cockburn, J.S., King, H.P.F., and McDonnell K.G.T. (eds), 1969. 'Private Education from the Sixteenth Century: Developments from the 16th to the early 19th century', in *A History of the County of Middlesex*, Vol. 1: *Physique, Archaeology, Domesday, Ecclesiastical Organization, The Jews, Religious Houses, Education of Working Classes to 1870.* London: British History Online: http://www.british-history.ac.uk/report .aspx?compid=22124&strquery=Hoxton

Colman, George and Garrick, David, 1766. *The Clandestine Marriage.* T.H. Lacy, London.

Cooke, William, 1835. *A Brief Memoir of Sir William Blizard, Knt.* Longman, Rees, Orme, Brown, and Co., London.

Corner, George and Goodwin, Willard, 1953. 'Benjamin Franklin's bladder stone', *Journal of the History of Medicine* (8) 4, pp. 363–77.

Credner, H., 1891. *Elemente der Geologie*. W. Englemann, Leipzig.

Da Vinci, Leonardo, translated by E. MacCurdy, 1938. 'Physical Geography', in: *The Notebooks of Leonardo da Vinci*, Vol. 1.

Davis, Michael T., 1995. *Behold the Man: the life, times and circle of Daniel Isaac Eaton, 1753–1814.* Unpublished PhD thesis, University of Queensland, Australia.

Davis, Michael T., 2004. 'London Corresponding Society (*act.* 1792–1799)', *Oxford Dictionary of National Biography*, Oxford University Press, online edn, January 2008.

Davis, Michael T. (ed.), 2001. *The London Corresponding Society 1792–99* (6 vols). Pickering & Chatto, London.

De Luc, J.-A., 1779. *Lettres Physics et Morales sur l'Histoire de la Terre et de l'Homme. Adressées à la Reine de la Grande Bretagne*. The De Tune, Hague, Paris.

De Vera, Mary, et al., 2008. 'Gout and the Risk of Parkinson's Disease: a Cohort Study' *Arthritis & Rheumatism* (59) 11, pp. 1,549–54.

Dodd, William, 1841. *A Narrative of the Experience and Sufferings of William Dodd, a Factory Cripple. Written by himself.* L. & G. Seeley, London.

Dunn, Peter, 2000. 'Dr. William Buchan (1729–1805) and his Domestic Medicine', *Archives of Disease in Childhood: fetal and neonatal edition* (83), F71–F73.

Eaton, Daniel Isaac (ed.), 1793–5. *Politics for the People.* Printed for D.I. Eaton, at the Cock and Hog-Trough, Newgate Street, London.

Eaton, Daniel Isaac, 1793a. *The Trial of Daniel Isaac Eaton, before Lloyd Lord Kenyon, and a special jury, in the Court of King's Bench, Guildhall, London, July the tenth, 1793; for selling a supposed libel, A letter, addressed to the addressers. By Thomas Paine.* Printed and published by the defendant, Daniel Isaac Eaton; aud [*sic*] sold by H.D. Symonds; James Ridgway; Rio [*sic*] Rickman; T.W. Hawkins; I.S. Jordan; J. Gale, Sheffield; R. Phillips, Leicester; Benjamin Flowers, Cambridge; Messrs. Berry and Robinson, Edinburgh; Mash and Read, Glasgow.

Eaton, Daniel Isaac, 1793b. *The Proceedings, on the trial of Daniel Isaac Eaton, upon an indictment, for selling a supposed libel, "the second part of the Rights of man, combining principle and practice" by Thomas Paine. At Justice Hall, in the Old Bailey. Before the Recorder of London. On Monday, the third day June, 1793.* Printed and published by the defendant, Daniel Isaac Eaton, No. 81, Bishopsgate Street, London. And sold by James Ridgway, York Street, St James's; H.D. Symonds, Paternoster Row; And all other Booksellers in Town and Country.

Eaton, Daniel Isaac, 1812. *Address of D.I. Eaton, Now Under Sentence of Eighteen Months Imprisonment in Newgate.* Printed by D.I. Eaton, London.

Elliotson, John, 1831. 'Paralysis agitans', *The Lancet London: A Journal of British and Foreign Medicine, Surgery, Obstetrics, Physiology, Chemistry, Pharmacology, Public Health and News* (1), p. 290.

Ellis, Henry, 1798. *The History and Antiquities of the Parish of*

Saint Leonard Shoreditch and Liberty of Norton Folgate in the Suburbs of London. J. Nicholls, London.

Emsley, Clive, 1979. 'The Home Office and its Sources of Information and Investigation 1791–1801', *The English Historical Review* (94), No. 372.

Farey, John, 1811. *General View of the Agricultural and Minerals of Derbyshire*. Vol. 1. McMillan, London.

Fauvell, D. and Simpson, I., *The History of British Winters* at netweather.tv: http://www.netweather.tv/index.cgi?action =winter-history;sess

Fleming, Caleb D.D., 1773. *A Dissertation upon the Unnatural Crime of Self-Murder: occasioned by the many late instances of suicide in this city, &c.* . . . Edward and Charles Dilly, London.

Ford, J.M.T., 1987. 'A Medical Student at St Thomas's Hospital, 1801–1802. The Weekes Family Letters', *Medical History*, Supplement No. 7. Wellcome Institute for the History of Medicine, London.

Gardner Thorpe, C., 1987. *James Parkinson 1755–1824*. Printed by A. Wheaton and Co. Ltd, Exeter. (Contains a facsimile of Parkinson's *Essay on the Shaking Palsy.*)

Gilliland, J., 2004. 'Mitford, John (1782–1831)', *Oxford Dictionary of National Biography*, Oxford University Press.

Goetz, Christopher G., 1986. 'Charcot on Parkinson's Disease', *Movement Disorders* (1) 1, pp. 27–32.

Goetz, Christopher G., 2011. *The History of Parkinson's Disease: Early Clinical Descriptions and Neurological Therapies*. Cold Spring Harbor Perspectives In Medicine, 1:a008862.

Grosley, J.P., 1772. *A Tour to London, or new observations on England*. Printed for Lockyer Davies, London.

Gurney, Joseph, 1796. *The Trial of Robert Thomas Crossfield for high treason*. Sold by Martha Gurney, bookseller, Holborn-Hill, London.

Hardy, Thomas, 1832. *Memoir of Thomas Hardy, founder of, and Secretary to, the London Corresponding Society, for diffusing useful political knowledge among the people of Great Britain and Ireland, etc. Written by himself. [With a preface by D. Macpherson.]* James Ridgeway, London.

Harland-Jacobs, Jessica, 2010. 'Freemasons and the political culture of the British Atlantic world, 1717–1798', *Hibiscus Masonic Review* (3), pp. 40–65.

Hawes, W. (ed.), 1774. *Transactions of the Royal Humane Society from 1774 to 1784*. Jno. Nichols, London.

Hay, Douglas, 2004. 'Kenyon, Lloyd, first Baron Kenyon (1732–1802)', *Oxford Dictionary of National Biography*, Oxford University Press, online edn, October 2009.

Heringman, Noah, 2004. *Romantic Rocks, Aesthetic Geology*. Ithaca and London: Cornell University Press, USA.

Higgins, John Robert, 2011. 'Fossil poetry, the birth of geology, and the Romantic imagination, 1790–1860', *Theses and Dissertations*. Paper 1073. http://scholarcommons.sc.edu/etd/1073

Holloway, S.W.F., 1966. 'The Apothecaries' Act, 1815: a reinterpretation', *Medical History* (10) 2, pp. 107–29.

Home, Everard. 1790. 'An Account of a Child with a Double Head. In a Letter to John Hunter, Esq. F.R.S.', *Philosophical Transactions of the Royal Society of London* (80), pp. 296–305.

House of Commons, 1816. *The First Annual Report on Madhouses*. W.I. Clements, London.

Hunter, John, 1859. *Observation and Reflections on Geology: Intended to Serve as an introduction to his collection of Extraneous Fossils.* London: Taylor and Francis.

Jenner, E. 1798. *An Inquiry Into the Causes and Effects of the Variolæ Vaccinæ, Or Cow-Pox.* London: Printed, for the author, by Sampson Low.

Kaufman, M.H., 2001. *Surgeons at War: medical arrangements for the treatment of the sick and wounded in the British Army during the late 18th and 19th centuries.* Greenwood Press, Westport, CT.

Klein, Ursula, 2004. 'Not a pure science: Chemistry in the 18th and 19th centuries', *Science* (306), pp. 981–2

Knapp, Andrew and Baldwin, William, 1810. *The Newgate Calendar: comprising interesting memoirs of the most notorious characters who have been convicted of outrages on the laws of England since the commencement of the eighteenth century;*

with occasional anecdotes and observations, speeches, confessions, and last exclamations of sufferers. Vol. 2. J. Robins and Co., London.

Kohn, George Childs, 2007. 'English typhus epidemic, 1837–1838', *Encyclopedia of Plague and Pestilence*, Third Edition. Facts On File, Inc., New York.

Lamarck, Jean-Baptiste, 1809. *Philosophie Zoologique*. Paris: Dentu et L'Auteur.

Lawrence, S.C., 1996. *Charitable Knowledge. Hospital Pupils and Practitioners in Eighteenth Century London*. Cambridge University Press, Cambridge. Chapter 1.

Lemaitre, Paul, 1795. *High Treason!! Narrative of the arrest, examinations before the Privy Council, and imprisonment of P.T. Lemaitre, accused of being a party in the Pop-Gun plot, or, pretended plot to kill the King!* J. Smith, at the Pop-Gun, Portsmouth Street, Lincoln's Inn Fields.

Lewis, Cherry, 2000. *The Dating Game: one man's search for the age of the Earth*. Cambridge University Press, Cambridge.

Lewis, Cherry, 2009a. 'Doctoring geology: the medical origins of the Geological Society', in: Lewis, C.L.E. & Knell, S.J. (eds), *The Making of the Geological Society of London*. The Geological Society, London, Special Publications, 317, pp. 49–92.

Lewis, C.L.E., 2009b. 'Our favourite science: the British Prime Minister and the apothecary surgeon, searching for a Theory of the Earth', in: Kölbl-Ebert, M. (ed.), *Geology and Religion: Historical Views of an Intense Relationship between Harmony and Hostility*. The Geological Society, London, Special Publications, 310, pp. 111–26.

Lewis, C.L.E., 2013. 'James Parkinson's "system of successive creations"', in: Duffin, C.J., Moody, R.T.J. & Gardner-Thorpe, C. (eds), *A History of Geology and Medicine*. Geological Society, London, Special Publications, 375, pp. 339–48.

Lewis, C.L.E., 2016. 'David Mushet, John Farey, and William Smith, geologising in the Forest of Dean', *Earth Sciences History* (35) 1, pp. 167–96.

Lewis, C.L.E., Green, P.F., Carter, A. and Hurford, A.J., 1992.

'Elevated K/T palaeotemperatures throughout Northern England: three kilometres of Tertiary exhumation?', *Earth and Planetary Science Letters* (112), pp. 131–45.

London Corresponding Society, 1792. *The London Corresponding Societies. Addresses and resolutions (reprinted).* London.

London Corresponding Society, 1795. *A Summary of the Duties of Citizenship!* Printed and sold by D.I. Eaton, No. 74, Newgate-Street; G. Riebau, 439, Strand; J. Smith, Portsmouth-Street; etc. London.

Lovat-Fraser, J.A., 1932. *Erskine*. Cambridge University Press, Cambridge.

Macfarlane, Alan, 1986. *Marriage and Love in England: modes of reproduction 1300–1840*. Blackwell, London.

Macilwain, George, 1856. *Memoirs of John Abernethy, with a view of his lectures, writings, and character*. Hatchard and Company, London.

Mackenzie, Henry, 1967. *Letters to Elizabeth Rose of Kilravock on Literature, Events and People 1768–1815*. Edited by Horst Drescher. London: Oliver and Boyd.

Maclachlan, Daniel, 1863. *A Practical Treatise on the Diseases and Infirmities of Advanced Life*. John Churchill and Son, London.

Maddocks, James and Blizard, William, 1783. *An Address to the Friends of the London Hospital, and of Medical Learning*. [No publication details given].

Maiden, William, 1812. *An Account of a Case of Recovery, After an Extraordinary Accident, by which the shaft of a chaise had been forced through the thorax*. T. Bayley, London.

Mantell, Gideon, 1846. 'A few notes on the prices of fossils', *London Geological Journal* (1) pp. 13–17.

Mantell, Gideon, 1850. *A Pictorial Atlas of Fossil Remains*. H.G. Bohn, London.

Mason Good, John, 1795. *The History of Medicine, so far as it relates to the profession of the apothecary*. Printed for C. Dilly, in the Poultry, London.

McCue, Daniel, 1974. *Daniel Isaac Eaton and 'Politics for the People'*. Ph.D. dissertation, Columbia University, USA.

McCue, Daniel Lawrence Jr, 2004. 'Eaton, Daniel Isaac (*bap.* 1753, *d.* 1814)', *Oxford Dictionary of National Biography*, Oxford University Press, online edn, October 2007.
McMenemey, W.H., 1955a. 'James Parkinson 1755–1824: A biographical essay', in: Critchley, M. (ed.), *James Parkinson (1755–1824)*. London: Macmillan, pp. 1–143.
McMenemey, W.H., 1955b. 'A note on James Parkinson as a reformer of the Lunacy Acts', *Proceedings of the Royal Society of Medicine* (48) 8, pp. 593–4.
Miele, Christopher, 1993. *Hoxton. Architecture and History Over Five Centuries*. A Hackney Society Publication, London.
Miller, J.S., 1821. *A Natural History of the Crinoidea or Lily-shaped Animals*. C. Frost, Bristol.
Mitford, John, c. 1825. *A Description of the Crimes and Horrors of the Interior of Warburton's Private Madhouse at Hoxton, Commonly Called Whitmore House*. Benbow, London.
Moore, Wendy, 2005. *The Knife Man: Blood, Body-snatching and the Birth of Modern Surgery*. Bantam Books, London.
Morgan, C.N., 1968. 'Surgery and surgeons in 18th century London', *Annals of The Royal College of Surgeons of England* (42) 1, pp. 1–37.
Morris, A.D., 1974. 'Samuel Dale (1659–1739), Physician and Geologist', *Proceedings of the Royal Society of Medicine* (67), pp. 120–4.
Morris, A.D., 1989. *James Parkinson, His Life and Times*. Birkhauser, Boston, Basel and Berlin.
Murphy, Elaine, 2001. 'The Mad-House Keepers of East London', *History Today* (51) 9.
Murphy, Elaine, 2004. 'A Mad House Transformed: The Lives and Work of Charles James Beverly FRS (1788–1868) and John Warburton MD FRS (1795–1847)', *Notes and Records of the Royal Society of London* (58) 3.
Newton, John, 1795. *The Trial at Large of John Thelwall. For high treason; before the special commission at the Sessions-House in the Old-Bailey: began on Monday, December 1, and continued until Friday 5, 1794*. H.D. Symonds, London.
Nicholson, William, 1809. *British Encyclopaedia, or Dictionary of the Arts and Sciences*. Longman, Hirst, Rees & Orme, London.

Norris Brewer, J., 1816. *The Beauties of England and Wales: Middlesex*. J. Harris, Longman and Co., London.

O'Connor, R., 2007. 'Young-Earth creationists in early 19th-century Britain? Towards a reassessment of the "scriptural geology"', *History of Science* (45), pp. 357–403

Parkes, S., 1815. *Chemical Essays*. Vol. V. Baldwin, Cradock, and Joy, London.

Pearn, J. and Gardner Thorpe, C., 2001. 'James Parkinson (1755–1824): A pioneer of child care', *Journal of Paediatric Child Health* (37) 1, pp. 9–13.

Pearson, George, 1798. 'Experiments and observations, tending to show the composition and properties of urinary concretions', *Philosophical Transactions of the Royal Society of London* (88), pp. 15–46.

Pollak, Pierre & Krack, Paul, 2007. 'Deep-brain stimulation for movement disorders: historical notes and personal remarks', in: Jankovich, Joseph and Tolosa, Eduardo (eds), *Parkinson's Disease and Movement Disorders*, Lippincott, Williams & Wilkins, Philadelphia USA, pp. 653–91.

Porter, D. and Porter, R., 1989. *Patient's Progress: doctors and doctoring in eighteenth-century England*. Polity Press, Oxford.

Porter, Roy, 1982. *English Society in the Eighteenth Century*. Pelican Books, London.

Porter, Roy, 2000. *London, A Social History*. Penguin Books, London.

Power, D'Arcy (ed.), 1886. *Memorials of the Craft of Surgery in England, from materials compiled by John Flint South*. Cassell & Co., London.

Priestley, Joseph, 1768. *An Essay on the First Principles of Government, and on the Nature of Political, Civil, and Religious Liberty*. J. Johnson, London.

Quist, George, 1979. 'Hunterian Oration: Some controversial aspects of John Hunter's life and work', *Annals of the Royal College of Surgeons of England* (61), pp. 381–4.

Rappaport, Rhodda, 1997. *When Geologists Were Historians, 1655–1750*. Cornell University Press, New York.

Reiter, Paul, 2000. 'From Shakespeare to Defoe: Malaria in England in the Little Ice Age', *Emerging Infectious Diseases* (6) 1, pp. 1–11.

Roberts, Shirley, 1997. *James Parkinson, 1755–1824: from apothecary to general practitioner*. Royal Society of Medicine Press Ltd, London.

Roe, Nicholas, 2004. 'Thelwall, John (1764–1834)', *Oxford Dictionary of National Biography*, Oxford University Press, online edn, September 2012.

Rogers, John, 1815. *A Statement of the Cruelties, Abuses, and Frauds, Which Are Practised in Madhouses.* Printed for the author, London.

Rowntree, Leonard G., 1912. 'James Parkinson', *Bulletin of the Johns Hopkins Hospital* (23), pp. 33–45.

Rudwick, M., 1997. *George Cuvier, Fossil Bones, and Geological Catastrophes*. University of Chicago Press, Chicago.

Sanders, W.R., 1865. 'Case of an unusual form of nervous disease, dystaxia or pseudo-paralysis agitans, with remarks', *Edinburgh Medical Journal* (10), pp. 987–97.

Select Committee, 1833. *Report on Factory Children's Labour.* Parliamentary Papers 1831–32, Vol. XV.

Sharpe, Pamela, 1998. *Women's Work: The English Experience 1650–1914*. Bloomsbury, USA.

Sheppard, F., 1998. *London, A History*. Oxford University Press, Oxford.

Sigsworth, E.M. and Swan, P., 1982. 'An eighteenth-century surgeon and apothecary: William Elmhirst (1721–1773)', *Medical History* (26), pp. 191–8.

Simond, Louis, 1815. *Journal of a Tour and Residence in Great Britain, During the Years 1810 and 1811*, Vol. 1. Eastburn, Kirk and Co., USA.

Smith, John, 1795. *Assassination of the King! The conspirators exposed, or, an account of the apprehension, treatment in prison, and repeated examinations before the Privy Council, of John Smith and George Higgins, on a charge of High Treason.* J. Smith, at the Pop-Gun, Portsmouth Street, Lincoln's Inn Fields.

Smollett, Tobias, 1771. 'The expedition of Henry Clinker', in: Scott, Sir Walter (ed.), 1835, *The Selected Works of Tobias Smollett in two volumes with the memoir of the life and writings of the author.* Carey, Lea and Blanchard, Philadelphia. Vol. II, pp. 309–468.

Sterne, Lawrence, 1768. *A Sentimental Journey.* Oxford: Oxford University Press.

Stott, Rebecca, 2012. *Darwin's Ghosts: in search of the first evolutionists.* Bloomsbury Publishing plc, London.

Stott, Simon, 2014. 'The wrong James Parkinson', *Practical Neurology*, published Online First: http://pn.bmj.com/content/early/2015/01/05/practneurol-2014-001043.extract

Stuart Jones, E.H., 1950. *The Last Invasion of Britain.* University of Wales Press, Cardiff.

Suzuki, Akihito, 2004. 'Burrows, George Man (1771–1846)', *Oxford Dictionary of National Biography*, Oxford University Press, Online edn, May 2007.

Thackray, John, 1976. 'James Parkinson's "Organic Remains"', *Journal of the Society for the Bibliography of Natural History*, 7 (4), pp. 451–66.

Thale, Mary (ed.), 1983. *Selections from the Papers of the London Corresponding Society 1792–1799.* Cambridge University Press.

Thelwall, John. 1793. *An Essay Towards a Definition of Animal Vitality: read at the theatre, Guy's Hospital, January 26, 1793; in which several of the opinions of the celebrated John Hunter are examined and controverted.* T. Rickaby, London.

Torrens, H.S., 2008. 'Parkinson, James (*bap.* 1730–1813)', *Oxford Dictionary of National Biography*, Oxford University Press, online edn, January 2008.

Tunbridge, P., 1971. 'Jean André Deluc, FRS, 1727–1817', *Notes and Records of the Royal Society* (26), pp. 15–33.

Virdi-Dhesi, Jaipreet, 2010. *The Criminalized Body II.* Online publication: http://jaivirdi.wordpress.com/2010/09/20/monday-series-the-criminalized-body-ii

Watkins, J. & Schoberl, F., 1816. *A Biographical Dictionary of the Living Authors of Great Britain and Ireland.* Henry Colburn, London.

Watson, Thomas, 1855. *Lectures on the Principles and Practice of Physic Delivered at Kings College London*. Lea and Blanchard, Philadelphia. Lecture XXXVIII.

Welch, W., 1822. *Religiosa Philosophia; or a new theory of the Earth in unison with the Mosaic account of Creation with an appendix on the plurality of inhabited worlds.* Printed for the author by W. Byers, Plymouth-Dock.

White, Gilbert, 1880. *The Natural History and Antiquities of Selbourne in the County of Southampton*. Bickers and Son, London.

Whitehouse, Tessa, 2011. 'Hoxton Academy (1764–1785)', *Dissenting Academies Online: Database and Encyclopedia*, Dr Williams's Centre for Dissenting Studies: http://dissacad.english.qmul.ac.uk/sample1.php?contid=6&detail=academies¶meter=showarticles&acadid=74#tabs-2

Wilkes, Sue, 2016. *Regency Spies: Secret Histories of Britain's Rebels and Revolutionaries*. Pen and Sword History, Barnsley.

Wollaston, William, 1797. 'On gouty and urinary concretions', *Philosophical Transactions of the Royal Society of London* (87), pp. 386–400.

Wood, S., 1960. 'Mr. Tipple's chest wound', *Medical History* (4) 3, pp. 210–17.

PICTURE CREDITS

Figures in the text

Page 5: St Leonard's Church, Shoreditch. From *'Church Bells', an Album of Notable Churches*, 1891, Church Bells Office, London. Courtesy: Bob Speel.

Page 17: An artist's impression of the new London Hospital, Whitechapel, in 1752. University of Bristol Library Special Collections.

Page 20: 'The body of a MURDERER exposed in the Theatre of the Surgeons' Hall', *Newgate Calendar*, 1794.

Page 28: The ward of the Receiving House of the Royal Humane Society in Hyde Park. *Illustrated London News*, 31 August 1844, p.144.

Page 30: A Royal Humane Society dinner at the Freemasons' Hall, London. Fry Collection, University of Bristol Library Special Collections.

Page 48: Sketches of the child with two heads, and its skull. Home, 1790.

Page 58: J. Sharpe's humorous print of 'The swinish multitude', circa 1793.

Page 61: Daniel Isaac Eaton, from *The Trial of Daniel Isaac Eaton, for publishing the third and last part of Paine's Age of Reason* (1812).

Page 76: A sketch of the brass tube from which the 'pop gun' was to be made. The National Archives (Ref: KB33/6/3).

Page 84: An engraving of the Privy Council, from around the time Parkinson would have stood before them. National Archives and Records Administration, via Wikimedia.

Page 124: The five stages in making the truss designed by Parkinson. Wellcome Library, London.

Page 132: James Parkinson's dissecting microscope which he gave to Edward Jenner in 1808. Photograph courtesy of Christopher Gardner Thorpe.

Page 162: The fossil tubipore in limestone from Derbyshire (centre) on which Parkinson performed his experiment. *Organic Remains of a Former World*, Vol. 2, Plate I. University of Bristol Library Special Collections.

Page 171: William Norris in Bethlem. Wellcome Library, London.

Page 182: The fossilised skeleton of the extinct megatherium. Parkinson, 1811c, Plate XXII, Fig. 1. University of Bristol Library Special Collections.

Page 188: Illustration from the cover of *The Villager's Friend*. Wellcome Library, London.

Page 198: Sketch of the chaise shaft that pierced Mr Tipple. Wellcome Library, London.

Page 200: William Clift's drawing of Mr Tipple's rib cage. Wellcome Library, London.

Page 218: Drawing from Charcot's lesson given on 12 June 1888, which demonstrates the typical flexed posture of Parkinson's disease.

Page 230: Sir William Blizard when he was Professor of Anatomy and Surgery to the Royal College of Surgeons. Wellcome Library, London.

Page 234: John Hunter's Museum at the Royal College of Surgeons, 1853. © Hunterian Museum at the Royal College of Surgeons.

Page 237: The 'exquisite' fossilised sea urchin, Cidaris, that was in Parkinson's collection. Parkinson, 1811c, Plate I, Fig. 5. University of Bristol Library Special Collections.

Page 242: The Jurassic ammonite *Parkinsonia parkinsoni*. Credner, 1891, p. 605.

Page 245: James Keys Parkinson. Courtesy John Reynolds.

Page 246: Sarah Anne Parkinson (née Warren) with children James Warren, Emma, Rosina and Margaret. Courtesy John Reynolds.

Page 248: The sailing ship *May Queen* in Nelson circa 1880. Tyree Studio: Negatives of Nelson and Marlborough districts. Alexander Turnbull Library, Wellington, New Zealand.

Page 252: Catherine Parkinson (née Cooke) with her grandson Frank Alfred Warrington. Courtesy John Reynolds.

Page 253 (top): Frank Warren Parkinson and his younger brother Henry.

Page 253 (bottom): Three of James Warren Parkinson's daughters: Sarah, Margaret and Susan Catherine. Courtesy John Reynolds.

Page 255: John Reynolds in Christchurch, December 2014.

Plate section

Plate 1: 'The London Corresponding Society alarm'd' by James Gillray, 1798. National Portrait Gallery.

Plate 2 (top): 'The Gout' by James Gillray, 1799. Wellcome Library, London. (bottom): Illustrations of ammonites from *Organic Remains of a Former World*, Vol. 3, Plate IX. University of Bristol Library Special Collections.

Plate 3: Illustration of a crinoid that Parkinson called the Stone Lily or Lily encrinite. *Organic Remains of a Former World*, Vol. 2, Plate XIV. University of Bristol Library Special Collections.

Plate 4 (top): The silicified sponge, *Chenendopora michelinii* Hinde, which illustrates the frontispiece to Parkinson's second (1808) volume of *Organic Remains*. (middle): The same image recoloured by Sowerby in Mantell's *Pictorial Atlas of Fossil Remains*, 1850. University of Bristol Library Special Collections. (bottom): A photograph of the actual specimen that can still be seen in the Natural History Museum. © The Trustees of the Natural History Museum, London.

INDEX